Turbo Flow

Using Plan for Every Part (PFEP) to Turbo Charge Your Supply Chain

Turbo Flow

Using Plan for Every Part (PFEP)
to Turbo Charge Your Supply Chain

Tim Conrad
Robyn Rooks

CRC Press
Taylor & Francis Group
Boca Raton London New York

CRC Press is an imprint of the
Taylor & Francis Group, an **informa** business

A PRODUCTIVITY PRESS BOOK

Contents

Foreword

I am amazed at how long it has taken even the most enthusiastic supporters of Lean to take Lean supply chain management seriously. It is certainly right to begin at home before preaching to others. And it is easier to do Lean within your factory walls or business unit. However, so many of the gains made there are simply lost when you look at the disconnections all the way up and down the supply chain. Hurrying up to wait does not get anyone anywhere! But I am encouraged that the time has finally come when more and more organizations are taking Lean supply chains seriously.

This makes this book very timely. It goes to the heart of Lean supply chain management—showing in vivid detail how to create a Plan for Every Part (PFEP)—detailing where it should come from and go to and exactly what should happen to it at every step on its journey. This is the foundation for any pull system and the basis for linking actions up and down the supply chain. It is the natural complement to the *Making Materials Flow* workbook on internal logistics published by the Lean Enterprise Institute several years ago. Supply chain management is about looking at the big picture from raw materials to the end customer. But it is also all about the detail of every movement of every part in the system. One without the other does not work.

Tim Conrad and Robyn Rooks are extremely well qualified to guide you through this journey. Tim and Robyn spent 18 years in Kentucky at Toyota's biggest U.S. plant where they were responsible for inventory management and planning, among many other things. We met when we were both helping another company create big savings by Leaning its supply chain. I was impressed by their rigor and patience in establishing a Plan for Every Part. Follow this book carefully and your patience will also be rewarded by very significant improvements in the performance and responsiveness of your supply chains. And give this book to your upstream suppliers and down-stream customers so they can mirror your learning. Good luck on your Lean journey.

Daniel T. Jones
Chairman of the Lean Enterprise Academy in the United Kingdom and co-author of The Machine that Changed the World, Lean Thinking, and Lean Solutions

Introduction

Does your company have extra money that you need to tie up in inventories? Do you have a lot of storage room for extra parts just in case you need to build something? Do you feel that making your material manager's job more comfortable should determine your inventory levels? Do you have extra production capacity that would not absorb overhead if you did not build *something*? If you answered yes to all of these questions, then this book is not for you.

On the other hand, if like most of us, you would rather invest your company's money in something other than inventory, then this book can help you. It is not for the faint-hearted and it is not for those who are looking for a shortcut on their Lean journey. Rather, it is for those who are willing to fundamentally change the way they do business.

"Our way of thinking is very difficult to copy or even to understand."

Fujio Cho
Chairman
Toyota Motor Corporation, 2007

The authors are often asked: What is the secret to Lean? Like Toyota we are not afraid to share our knowledge, but like Toyota we know most companies lack the courage to take the steps necessary to complete this Lean journey. It will take courage to move away from fixed absorption, Material Requirements Planning (MRP), and all of the other scientific and financial measurement tools we have built over the past 60 years. It will take courage to simplify your business and adopt the Toyota model of Just-in-Time (JIT). It will take courage to install the disciplines of Lean manufacturing.

The truth is, emulating Toyota's success is more than just adopting a few tools. Toyota created a system, the Toyota Production System (TPS), that is more than the sum of the tools. It is a process of strategy deployment that drives every team member in the organization to utilize the tools of TPS to meet the strategic needs of the company. The one question that was asked of every proposal at Toyota was a simple one: "What is best for the business?" The concept here was that Toyota looks forward to make its gains; they do not rely on past history, except to learn from their failures. It is OK to fail; this is the way we learn, whether at Toyota or in everyday life. This is a different way of thinking and the courage comes into play where most companies make decisions to react: knee-jerk, firefight, or just survive. Toyota, on the other hand, plans ahead: it is not unusual to have a 10-year plan for a facility consisting of everything from expansion plans to market share to janitorial services cost.

Giving Everyone Involved the Courage to Make It Work

Before moving on to the content of this book, we would like to discuss in more detail the courage required by everyone involved. When you start down the path toward a Lean culture you are going to run into at least three mindset types:

1. Optimist: Those who are willing and able to see where your vision will take the company and are on board to support you and do "what is best for the business." These are the people who typically have a good understanding of Lean from your training, another employer, or schooling.
2. On the fence: Those who are a little on the "I don't know" side; they want to believe but don't have a good understanding of what Lean truly is. They want to believe that change can be made without losing their jobs, but there have been so many "flavors of the month" introduced over the years that they will sit back and see how things play out: "We'll just have to wait and see; I'll do what they ask."
3. Pessimist: "This will never work," "It's been tried before," and, "All they want to do is take our jobs away and they need me because I'm the only one who can run that machine." These people are usually very vocal but usually not in an upward manner. This is gossip, and it has to be stopped before it ever gets started. This will be particularly difficult in an established facility where gossip has been allowed to go on for years.

Unfortunately, a lot of employees will see Lean as threatening. This way of thinking comes from the fact that most organizations do not give the proper training before going to the *gemba* (floor) to start an implementation. Another big mistake many companies make is not pulling in people from the floor to participate on the implementation teams. If you only allow management or engineers to implement Lean, then you are instilling a mindset that you are "doing it *to* us" versus "doing it *with* us."

When a problem or a negative is taken upward to any level and resolved, gossip is taken out of the equation. However, once a problem or a negative is discussed with a peer or below, you have the beginnings of gossip. Gossip will kill a Lean transformation, so you must have the courage to fix this problem.

The Goals of This Book

The authors spent a combined 18 years at Toyota Motor Manufacturing, Kentucky (TMMK). During this time we learned that inventory management at Toyota was more than just kanban, JIT, or daily milk runs in the logistics network. Inventory management begins with your parts structure and is tied to how and where the parts are used within your factory. How much inventory you need is based on how much you plan to build. This then determines the frequency of delivery, your standard packs, and ultimately how often you need to pick up from your supplier. Our *sensei* (teacher) used to say that inventory movement is the pacemaker process of the assembly line at Toyota.

Understanding this was a struggle until the realization came that inventory movement, or flow, is the foundation of JIT:

■ The right part
■ At the right time
■ In the quantity needed

If the Toyota Kentucky plant wanted to build more than was needed, it would run out of components. If the plant built out of sequence, the logistics would not provide the right parts at the right time. If the plant tried to build ahead, it would eventually run out of components. Every vehicle built at TMMK in Georgetown was sold to a customer—Toyota Motor Sales USA, the sales arm of Toyota Motor Corporation America—and every vehicle was scheduled for manufacturing based on when it was needed to line off the final assembly line. The line off date was determined by the promised delivery date, based on the logistics lead time to get that car on the dealer's lot.

All of this requires a detailed inventory plan. This is known as a Plan for Every Part (PFEP). In this book we teach you how to use your PFEP to manage your raw materials, work in process (WIP), and finished goods inventories.

Chapter 1

Toyota Practiced Lean before It Was Called "Lean"

Before we get to the details of how you can use Plan for Every Part (PFEP) to turbo-charge your supply chain, it is important for you to have a better understanding of what motivates a company to become Lean. This is best explained by looking at the origins of Lean at Toyota to gain a better understanding of your own motivations for pursuing Lean concepts and specifically PFEP and flow.

Origins of a New Idea

Toyota Motor Company (TMC) traces its beginnings to the Toyoda family business, the Toyoda Loom Works. Early on, the Toyoda family was not convinced that the future of the company was in automobiles. Early prototype vehicles were hardly the resounding success we see in today's marketplace. In fact, the first prototype trucks barely made it to the Tokyo auto show with their wheels intact. Military necessity, more than practical application, saved the company during World War II. Toyota trucks were the workhorse of the Japanese Army and helped the company survive the war years.

Postwar, the Japanese government adopted a highly structured economic planning model. For a company to obtain scarce capital for modernization, the company needed the blessing of the Ministry of Trade and Industry (MITI), the government bureaucracy that managed access to capital. MITI recommended that Toyota not continue as an automotive company. MITI felt that Mitsubishi's and Nissan's production capacity was sufficient to meet the needs of the Japanese market. However, the Toyoda family felt differently. Toyoda felt that there was a significant opportunity for a company willing to build vehicles to meet global needs, not just the needs of the local market. Because MITI blocked Toyota's access to capital, Toyota was forced to raise capital by first selling cars and then building them. The lack of cash forced them to think about flow, the goal being to get paid for the end product at the same time as or before paying for components.

Improving on the New Paradigm

"I really did have something in mind. To compare a Ford V-8 with a four-engine Liberator bomber was like matching a garage with a skyscraper, but despite their great differences *I knew the same fundamentals applied to high-volume production of both, the same as they would to an electric egg beater or to a wristwatch.*" *(Italics added)*

Charles E. Sorenson
My Forty Years with Ford. New York: W.W. Norton, 1956

With the Allied victories of World War II, Japanese industrialists began looking at the war effort and the massive quantities of materials and processes behind it. They started studying American production methods and had a keen interest in the way Henry Ford had taken a patch of land in Willow Run, Michigan, and turned it into the embodiment of American ingenuity and productivity. Charles E. Sorenson, Henry Ford's right-hand man, envisioned a plant that would change the way a B-24 Liberator bomber was designed and manufactured. This idea would not produce one bomber a day, as it was currently being done at a plant in San Diego, but rather would produce one B-24 bomber every hour. Using manufacturing techniques from the Ford production system, Sorenson made this vision a reality, knowing that the Germans had neither the facilities nor the conception for greater bomber mass production.

Sorenson submitted his plan, sketched on a napkin, to Edsel Ford over breakfast the morning after visiting the San Diego plant. Edsel agreed to his proposal and made plans to meet with Major Fleet to propose the building of a facility to support the Allied forces with B-24 bombers. At its peak, Willow

Run was producing 25 bombers in a 24-hour period, with 488,193 part numbers, 30,000 components, 24 major subassemblies, 25,000 initial engineering drawings, 10 model changes in six years, and thousands of running changes.

At Toyota, Taiichi Ohno and Shigeo Shingo studied these findings and began looking at the Ford production system for implementation. They began to implement these techniques into the Toyota production system. From the studies of Willow Run they saw that inventory played a very central role. With the cash restrictions on the company, it became clear that they had to rethink the way Toyota would support the manufacturing processes; they had to adapt the Ford system to be a truly Just-in-Time system. This meant rethinking the entire spectrum of the way planning was performed, processes were built, engineering was completed, and suppliers (internal and external) produced, packaged, and shipped components to the assembly floor.

For internal processes, Ohno suggested that Shingo begin work on reducing long lead times for changeovers to allow for smaller batch sizes to be produced. This had to change: Toyota could not continue, as many did, to run parts simply because the man or machine was available. They wanted to use the Ford model of an almost continuous flow of material but not with the large inventories. Shingo made this possible by focusing on the Single Minute Exchange of Dies (SMED) and running small batches on a concept of Every Part Every Day, which allowed Toyota to have the flexibility that Ford just didn't think he needed. Ford refused to change his system as yearly model changes, multiple colors, and wanted options came onto the scene; as these changes happened, the Ford system started to break down.

Another shift from the American models in thinking occurred during the years that General Douglas MacArthur was overseeing the occupation of postwar Japan. General MacArthur pushed for the labor unions, whereas Toyota, seeing the way the American labor model did not use the brain power of the worker, wanted to take advantage of the untapped potential the workforce had in controlling their destiny and making the processes better. By understanding what William Edwards Deming, Joseph M. Juran, and Kaoru Ishikawa had done with quality, Toyota used the Quality Circle to pull out those ideas by problem solving and started using a daily activity called kaizen (improvement) to utilize the workers' knowledge and continuously improve the process.

Thus was born the first pull system: manufacturing based not on what the company thought customers would buy, or what was most profitable to build, but rather on what customers actually bought. Toyota knew this build-to-order system would not work forever, so ultimately a dealer network was established to provide a minimal inventory for customers to buy and drive home immediately. However, limited capital kept the size of this inventory very low as dealers were still required to purchase the finished vehicles.

Moving on Toward PFEP

This also required a flexible manufacturing system that could easily be adapted to changes in customer demand for vehicles, specifications, options, and even color. Supporting this flexible system would require large inventories of raw materials. The vehicle combinations available to customers were virtually limitless, so the plant needed a way to signal to the suppliers what parts were needed for each car.

The Right Part, at the Right Time, in the Right Quantity

Investments in large inventories were out of the question in Toyota's factories. Instead, suppliers were invited to build plants nearby and Toyota would buy what the plant built, as long as the plant built based on information provided by Toyota and delivered the parts exactly in the quantity needed when Toyota needed the parts.

After visiting America, Taiichi Ohno devised the kanban (visual card/board) as a communication tool. To buy forgings for his machining shop at Toyota, these cards allowed for only a stated number of parts in a bin and assured the honor of the forging producer because only quality parts were allowed to be passed on to the customer. Ohno's inventory plan was based on daily usage rate, delivery frequency, and packaging size: the characteristics we now call Plan for Every Part.

Ohno's signal kanban (Figure 1.1) was similar to this and was a production instruction kanban. A signal kanban is a replenishment trigger that is normally used for large batch production runs, such as a stamping operation. The signal is set at a predetermined production lot size, so that when the inventory reaches the rack or container that contains the signal it triggers a production run. In Figure 1.2, the signal kanban is placed in a stack of flat steel blanks at a calculated number of pieces from the bottom of the stack. When blanks are consumed downward from the top of the stack, and the signal kanban is reached, the kanban is pulled and taken to the planner to signal, or trigger, the replenishment cycle to begin. The remaining blanks in the supermarket location after the signal kanban has been pulled are the

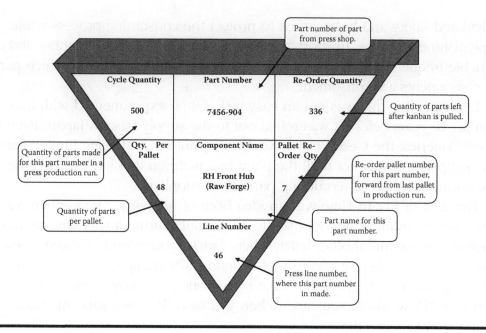

Figure 1.1   Signal kanban.

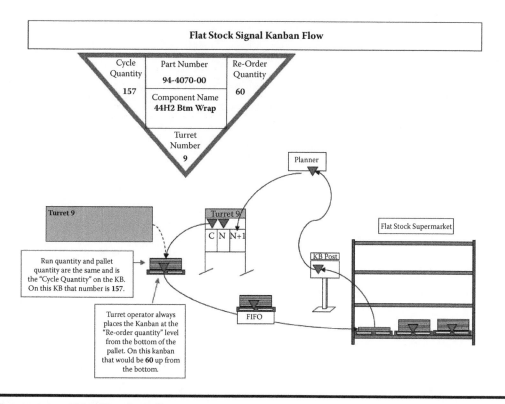

Figure 1.2   Signal kanban flow.

calculated safety and buffer stock to protect the consuming process while replenishment occurs. In short, the production run is a fixed quantity and a variable frequency, with the frequency determined by the rate at which parts or components are consumed.

Implementing this was not an easy task. Ohno experimented with kanban for 10 years before it was rolled out to the supply base in Japan. Even in North America, the Georgetown, Kentucky plant opened using limited kanban suppliers. It was not until the plant had matured that kanban became the normal method of reordering parts from suppliers.

The idea of Just-in-Time was an idea born of necessity. The kanban was a simple communication tool used to tell manufacturing it was time to make another part or run another batch where batch processing made sense. Even Toyota understands that you cannot economically stamp one part at a time, or stamp every part, every day. That is true one-piece flow. Ohno's instruction was, "Flow where you can." When you can't flow, use supermarkets (buffers) to facilitate flow.

Chapter 2

Understanding Plan for Every Part

Most companies beginning a Lean journey focus on the assembly process, without trying to link value creation to the entire supply chain. Figure 2.1 illustrates a simplified view of a typical supply chain. Orders come in to a company with unpredictable frequency. There is usually some kind of order management logic built into the enterprise system that releases orders in a logical sequence. Orders then get dispersed to various manufacturing sites. The manufacturing sites then release material requirements to their supply

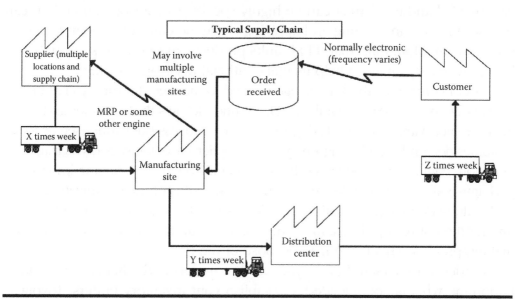

Figure 2.1 Typical supply chain.

base receiving components to build the order. Because multiple manufacturing sites are involved, the customer order goes to a Distribution Center (DC) where orders are consolidated and shipments are prepared and sent to customers. This lead time can be as long as five to six weeks or more.

Inventory Buffers Explained

Often inventory buffers are maintained in regional distribution centers to reduce customer fulfillment lead times. Orders go directly to the DC; then manufacturing signals are sent back to the plants from the DC. This does nothing to drive speed or efficiency into manufacturing cycles or lead times. Our experience teaches us that more waste (the waste of waiting for fulfillment and inventory) can be driven from the value stream by focusing on the entire supply chain as a whole, rather than by just focusing on the manufacturing or distribution pieces independently. To be successful, our metrics (how we measure success) must be looking at the entire value stream–supply chain, not individual pieces.

Plan for Every Part (PFEP) is the basis not just for inventory levels but also for packaging, storage racks, delivery frequency, routes, planning, and equipment layout. The entire design and layout of the plant, manufacturing cell, logistics network, and DC is driven by the PFEP. The PFEP is always unique to the value stream. This makes the design of the PFEP challenging as the data needed and the format can be highly specific to your operation. You can see the difficulty in creating and applying a generic concept to your specific value stream. More than that, PFEP links products and processes throughout the entire supply chain to help us make rational decisions quickly.

In most companies the knowledge needed to manage the supply chain is spread out and not readily available. Distribution knows the customer and what the customer wants. Manufacturing knows their process but cannot see or feel customer demand. Purchasing knows the supplier base, logistics owns transportation, engineering knows the Bill of Materials (BOM), packaging knows packaging: all of the knowledge is owned within silos of the organization.

A value-added supply chain consolidates this knowledge in one place, linking the entire supply chain to information that is critical to the decision-making process. This is the true purpose of the PFEP.

In order to understand PFEP you need to understand what is—and, more important, what is not—needed to establish your inventory buffers. Toyota teaches us to look at seven forms of waste. They acknowledge that there may be more than seven, but if you first focus on these seven, it will drive cost out of your supply chain. Eliminating the seven wastes will enable you to more effectively develop your inventory plan, your PFEP.

Understanding Waste

Think of waste as anything for which a customer is not willing to pay. Therefore, any waste activity is non-value-added. The customer expects you to make a widget, screw, and so on. He is willing to pay you for your time to produce his required product at a quality level he has specified and place it in some form of container that ensures that quality upon delivery. He is essentially buying a block of your time and nothing more. Therefore, he is buying value-added time only. Anything else you want to put into his product is fine, as long as it is at your cost; you can "waste" as much of your money as you want. Figure 2.2 shows the breakout of value-added, non-value-added waste, and incidental work. Each portion of the circle represents an element of the work motion required to produce an item.

Value-Add: Any activity that transforms or shapes raw material or information as required by a customer need

Non-Value-Add (Waste): Activities that consume time, resources, or space but do not contribute to satisfying customer need

The Toyota Production System (TPS) teaches us about eliminating, measuring, and problem solving the seven wastes:

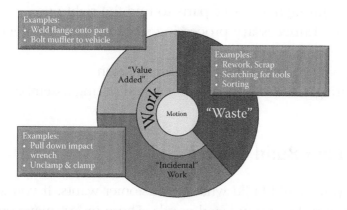

Figure 2.2   Work motion.

1. *Overproduction:* Producing over customer requirements, producing unnecessary materials/products

   Producing parts on Monday that are not shipped to the customer until Friday

   Producing parts that are not required just because the machines and people are available

2. *Waiting:* Time delays, idle time, machine watching

   Waiting for raw materials to be delivered

   Waiting for a machine to complete its cycle

3. *Conveyance:* Multiple handling, delay in material handling, unnecessary handling

   Moving parts to and from storage areas

   Transporting parts from the machining areas to the final assembly area

4. *Overprocessing:* Unnecessary processing steps or work elements/procedures

   Painting an automotive seat frame that is covered in cloth and the customer never sees

   Inspection, deburring, washing, and so on

5. *Inventory:* Holding or purchasing unnecessary raw supplies, work-in-process, finished goods

   Over 10 days of raw material supply

   A box containing 1,000 parts waiting to be assembled after being machined

6. *Motion:* Actions of people or equipment that do not add value to the product.

   Turning around to pick up a part

   Sorting through a box of parts to find the right one

7. *Correction:* Unnecessary processing steps or work elements/procedures

   Scrapping parts that failed final inspection

   Repairing a surface that was scratched during assembly

What Should I Build Today?

Like Toyota, you should build what the customer wants. If you are lucky, the customer issues orders for finished goods. These orders may come directly to a manufacturing or assembly plant or to a distribution center.

Orders coming directly to a manufacturing or assembly plant are fairly straightforward. Based on shipping lead times, you can build these products and move them directly to a loading dock where pallets are built, loads are built, products ship, and shipping confirmations are sent so that customers can track orders. Dell uses this model to build your computer to your specifications. When you go online (www.dell.com) you select what you want. You are asked first if you want a desktop or laptop. You are then asked how you will use your computer: home, home office, small business, or large business. This decision leads you to a product selection where you can begin to customize your computer. The basic model includes similar processor and memory configurations, basic graphics cards, monitor, printer, and so on from a starting price. You are then given the option to customize your system by upgrading monitors, printers, CPUs, memory, software, and so on. These customizations may add time to the delivery schedule but at the end of the order process you receive a series of e-mails with shipping times and tracking numbers. This creates a Do-It-Yourself (DIY) environment where you assume tracking responsibility for your shipment through the Dell logistics network. Customer service calls to Dell are reduced as you shoulder this responsibility yourself.

Most manufacturers (including Dell) keep inventory in some kind of distribution or service center. These centers are stocked with inventory based on historic sales trends or sales forecasts. (The Dell model keeps inventory in the unassembled or partially assembled stage.) These inventory buffers are also used to compensate for manufacturing inefficiencies (waste).

We cover up long lead times or difficult-to-manage changeovers (waste) by keeping large inventories to protect customer service targets. We also add some level of safety or "fluff" to this inventory to buffer against unexpected or unexplained demand (waste). We buffer for poor attendance (waste) because we can't rely on cross-training. We buffer to meet our budget and end-of-month variances (waste). All of these wastes are variations to the process in one form or another. Remember: variation not understood results in inventory; long internal or external lead time results in inventory. Some variations will always exist, but the goal is to understand and reduce the variation and thus reduce the need for inventory protection.

In the example shown in Figure 2.3, orders go directly to the DC, and the DC inventory triggers a replenishment order back to the manufacturing plant. To make this system work, inventory buffers would be required to cover the manufacturing and logistics lead times. By having these inventory buffers, you can reduce your order fulfillment lead time to days instead

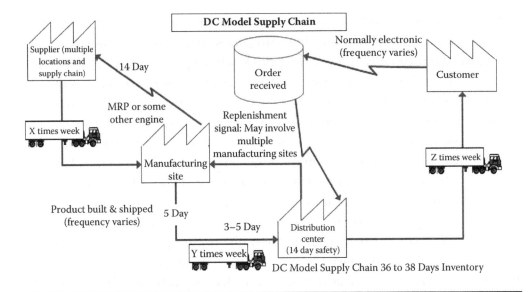

Figure 2.3   Supply chain example with distribution center.

of weeks. These inventory buffers are calculated based on average historic inventory trends; however, these historic inventory trends fail to cover variations in demand.

Waste in our supply chain causes us to create buffers to account for supply chain variation. This variation is caused by unreliable manufacturing performance, transportation systems, and unpredictable customer demand: all forms of waste.

How Much Inventory Do I Need?

When we ask companies why they are beginning a Lean journey, one of the goals they often lay out is inventory reduction. We want to "reduce inventories by $Xm." This is an admirable goal. It is also irrelevant.

It is irrelevant because when asked how much inventory is needed, the answers are usually based on total dollars tied up in inventory, or inventory turn targets, not on how much inventory is maintained at the item level. The inventory is usually based on a budgeted inventory level without regard to customer demand for end items. I have never seen customers place orders that say, "Send me $150,000 of inventory," allowing the supplier to randomly pick what is shipped. Customers order end items. We must manage end items.

PFEP forces you to make a rational inventory decision, at the item level, based on lot size, delivery frequency, manufacturing cycles, and customer demand. Inventory reduction is no longer relevant because inventory levels move in proportion to customer demand. When demand goes up, so will inventories. When demand drops, inventories will also drive down because the decision on inventory level is based on true need not on a forecast budget. True inventory reduction can only be achieved by reducing cycle times or rationalizing part number strategy.

The goal is to use the PFEP to develop an inventory plan throughout your value stream. Remember, inventory is there for nothing more than hiding, or buffering, inefficiencies (waste) within your process, regardless of whether that inefficiency is in lead time, machine, or manpower. Of these three types of inefficient waste, lead time and machine are the most costly to your organization. Leave the manpower part alone; it only accounts for 10 to 15% of your total cost in a product. Although manpower is the easiest place to cut, you will cut much more using the PFEP to find the true waste in the value stream flow.

Let's talk about the way we use inventory in a mass manufacturing environment. In good times, we tend to:

■ Hire manpower.
■ Not really care about machine utilization or maintenance condition.
■ Make promises to customers for unrealistic delivery dates.
■ Set targets for reducing operating cost by dropping a local supplier to save a nickel on a part by going to a low-cost country.
■ Consider inventory as an asset because it makes the organization look good to the shareholder.

In bad times we tend to:

■ Knee-jerk and start cutting the easy stuff first such as manpower.
■ Cut back on machine utilization and still don't focus on maintenance condition.
■ Call customers and offer fire sale prices on inventory.
■ Set inventory reduction activities into motion by buying no more raw material than you ship out today.
■ Issue harsh warnings to plant and production control managers about consequences of not getting inventory levels down.
■ Cut off suppliers with no ramp-up plan for bringing the processes back on line when things turn around.

This is the point in helping an organization that drives the development of a PFEP, because we measure the wrong things. Too often the one consideration taken in the bad times is "What do I have, and is any of it sellable?" The following is the series of events of one client who went through these unfortunate steps:

- The DC was not considered a customer of the supplying manufacturing plants, even though the customer service department, the link to the end customer, was located at the DC and was the true front line to the end customer. The DC was nothing more than a dumping ground for the plants to push inventories into, so you could get the product out the door, at the plant, and capture the variance on manpower and machine (meeting your budget), using the Manufacturing Resource Planning (MRP) system to pull ahead and chase the cumulative.
- The DC manager had no control or ownership of the inventory she was receiving. All inventory was owned by the supplying plants and during good times, no target for Days on Hand (DOH) was managed.
- Plant production control manipulated the MRP increase in production output to meet budget requirements, ignoring the customer demand fed by the DC customer service staff.

So how do Lean and the Toyota production system help you in becoming more flexible? TPS teaches us the key philosophies of process flexibility.

- An integrated system of product development, process engineering, operations management, and corporate management processes.
- A manufacturing philosophy that shortens the lead time to deliver high-quality, low-cost products through the elimination of waste.
- Create a system and philosophy to maximize and streamline the flow of information and material. This flow will produce a process that will provide the benefits shown in Figure 2.3. Flow will provide the true benefits of Lean, allowing us to do more faster, while tying up fewer resources. Figure 2.4 is from a Massachusetts Institute of Technology (MIT) study showing the gains a Lean flow system can generate for an organization.

Remember: Lead time equates to inventory and inventory equates to cash you have tied up in your system that cannot be readily accessible.

Lead time reduction at Toyota, or Just-in-Time, is what allows them to be so competitive. They use the cost reduction principle to gain a foothold on

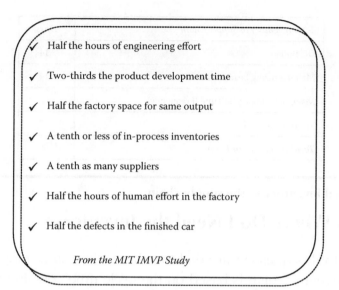

Half the hours of engineering effort

Two-thirds the product development time

Half the factory space for same output

A tenth or less of in-process inventories

A tenth as many suppliers

Half the hours of human effort in the factory

Half the defects in the finished car

From the MIT IMVP Study

Figure 2.4   The benefits of flow of information and material. (Adapted from Krafcik, J. F. *Sloan Management Review,* 30(1), fall, pp. 41–52, 1988.)

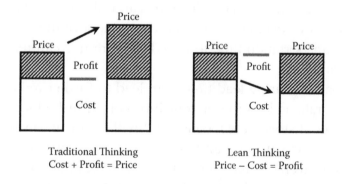

Traditional Thinking
Cost + Profit = Price

Lean Thinking
Price − Cost = Profit

Figure 2.5   Cost plus model.

the competition rather than the cost plus principle to accounting practices used by most companies (see Figure 2.5).

The cost plus principle means essentially that sales price equals profit plus costs. This means that when costs rise, the sales price must also rise in order to ensure a profit, regardless of whether the market can bear the price increase. This, then, places us at the mercy of forces outside the company.

Using the cost reduction principle, price minus cost (waste) equals profit. This is the Lean way of thinking. To maintain the same level of profit when cost increases, kaizen improvements are implemented by the entire work-force to remain competitive. Now, the company controls the forces that are working against it.

Total Inventory in Supply Chain	Days
DC Buffer	14
Manufacturing Lead Time	14
Manufacturing Cycle Time (Frequency)	5
Transportation Time	5
Total Inventory in Loop	38

Figure 2.6   Total inventory in the supply chain.

When and Where Do I Need the Inventory?

The answers to this question are tied together in inventory loops. Each element of the inventory plan is based on the cycle time contained in the loop. Movement triggers are based on consumption and the signal is a kanban that notifies the preceding process to replenish a supermarket or stores area. Figure 2.6 shows in greater detail the total inventory in the supply chain loop from Figure 2.3. Inventory reductions can only come when we reduce the number of days' coverage in a given component of the loop. For example, reducing transportation time to two days from five can reduce inventories by three days.

Reducing manufacturing lead times can lead to further reductions. Once you know the strategic location of your inventory (finished goods warehouse, raw material stores, etc.), your PFEP will lead you to your movement and replenishment frequencies and cycles as well as your buffers based on where you want inventory reductions to occur in your supply chain.

It is important to remember that this part of the inventory loop is driven by the lead time of administrative processes as well as physical movement.

PFEP will lead us to reducing cycle times in our value stream, helping us to get product to the customer faster. Waste elimination is the first step in reducing cost in our supply chain. Using these tools together and gaining an understanding of our value stream will help us create a supply chain that is ready to be turbo-charged.

Chapter 3

Management of PFEP

Pulling all the information required to build a Plan for Every Part (PFEP) into one location from the many sources that control it and changing to just one control point can be the most difficult part of the process. There are two keys to successfully creating a supply chain managed by PFEP: every part and ownership of the PFEP. When we begin to focus on the Bill of Materials (BOM) and match it to what we actually build through our value stream, it will become very obvious that we truly do not know as much about what we need to produce a product as we may think. We now begin to focus on our finished goods; our raw materials and their suppliers; how tooling, lubricants, and spare parts affect our process; and how takt time drives all of these items throughout the supply chain. We want to help you see the whole process and break down the silo's thinking so we understand the true importance of materials, and how they relate to getting a product from your suppliers' process through to the end customer.

Who Owns the PFEP?

Ownership of the PFEP is just as crucial in supporting the supply chain as knowing what components are needed to build an item. The PFEP is a living document and the DNA of your value stream(s). Therefore, maintenance and management of the PFEP is critical to ensure that there is one version of the truth. There will need to be one point of contact to control these changes, coordinate the timing, and document each change. We ask the people who currently control pieces of the PFEP to cede control to one person in order to create a sense of ownership and control of the PFEP. Anyone in the organization should be able to view, access, and print the PFEP or certain elements of it but not be able to alter, delete, or add items. We suggest the formation of a new role, PFEP coordinator. This person will be the point of contact to keep the information pure, current, and usable.

Every Part

Including every part means that you include everything: consumables, cardboard, dunnage, hardware: all of those items that we have historically abdicated trying to control.

Because of the overall total number of finished goods components, tooling, and so on, part numbers in their system covering all the products they produce and consume, many think that managing every part is unrealistic and not doable. PFEP is the tool that allows you to manage all components and finished good part numbers. Only by managing every part can we gain control and a deeper understanding of what makes our value stream work and, more important, what makes our value stream *not* work.

Breaking Down the "Every"

- *Finished Goods:* If you are a build-to-stock organization, you are manufacturing and holding stock in a finished goods market, whether in-house or at a DC, to support variation in your process and variation in customer demand. Finished goods inventories are there to buffer for inefficiencies throughout your value stream: capacity, long changeover, poor planning, and the like.
- *Raw Materials/Components:* This includes any item that is required in the manufacture of a customer finished part number. This area is where we will find the most opportunity for cost reduction. Below are some pitfalls we have to consider when we begin the PFEP:
  - Bill of material accuracy
  - Created phantoms within the BOM (phantoms are higher-level part numbers created by either accounting or engineering to reduce the number of trackable part numbers on a BOM within the Manufacturing Resource Planning [MRP] system. These are normally used where a subassembly is produced; sometimes the components of the subassembly are expensed part numbers, Vendor Managed Inventory [VMI], and other times they are just left off the BOM entirely but are still ordered,

stored, and used in your process. Either way, when you start to research part numbers to populate the PFEP, the components of the phantom will be omitted if you do not know and understand that they exist.)
- Items not on the BOM because they are considered expense items such as bags, boxes, packaging material, screws, bolts, washers, and so on
- Long lead times and high minimum order quantities on materials purchased from low-cost-country suppliers
- Suppliers who run their own trucks, schedule component delivery to your facility, and so on

■ *Lubricants and Spare Parts:* Any maintenance, repair and overhoul (MRO) item that is required to maintain the equipment in an ideal state to ensure zero downtime
■ *Tooling:* Tooling requirements need to be understood. Every part number needs to be listed within the PFEP and matched with its appropriate tooling. This is one area that will lead to high levels of inventory very quickly if you do not understand the tooling requirements. In Figure 3.1, you can see that we have very few dies, spread across multiple machines due to steel thickness and tonnage of the machine. This becomes clearer as we get into the section on dynamic capacity planning.

Toyota Cost Reduction Model

Toyota uses a simple management process to drive cost reduction. Decision making is pushed down to the lowest possible level in the organization to the group leader or team leader level. Cost reduction is the constant drive to eliminate the seven wastes. The measurements Toyota uses are

■ Seconds
■ Pennies
■ Tenths of a person

Toyota does not measure the value stream in days, hours, or minutes; the value-added and non-value-added times are expressed in seconds. Workload *Yamazumis* (Japanese term, applies to stack charts) are measured in tenths of a person, not in headcount; and cost is measured in pennies, not dollars. This allows small changes to drive improvement instead of looking only for major changes to drive down costs.

Some Lean practitioners will argue that this incremental improvement will not produce the dramatic improvement that a five-day, cross-functional

Die Number / Press Number	1.6	13.71	13.72	13.94	13.98	14.25
47		5	4			3
82		4	3			2
83						3
125		5	13			1
137						
149				1		
154				10	1	
158		2	11			4
796	4					
Part Numbers Requiring Die Number Per Machine	4	16	31	11	1	13
Number of Presses Die Required On	1	4	4	2	1	5
Total Number of Dies In-House	0	3	1	3	2	1
Short Dies	2	2	4	0	0	5

Figure 3.1   Die utilization.

kaizen event can drive. We believe that kaizen events can be very helpful in teaching how to go about improvement, but for truly sustainable change, it must happen at the smallest work elements of each process.

Based on the idea of measuring and making improvements at the lowest element possible, all components of the product must be measured and accounted for in the PFEP. Financial decision makers may have decided that it is easier to expense consumables than it is to manage them as part of the BOM, but we cannot manage the supply chain selectively. We have to manage it totally. More savings will be realized when we focus on pennies than when we focus on dollars.

Ownership of the PFEP

Silos of knowledge enable us to efficiently manage pieces of the supply chain. Purchasing specialists, logistics specialists, engineering specialists, and materials specialists all bring a level of expertise to their respective areas. Creating efficiencies within the silo often leads to the law of unintended consequences. Efficiency at one point adds cost at another. The goal is to eliminate waste not move it.

One company we worked with repackaged a finished part number from a single pack into a bulk pack for their customer in a regional DC. This added hours to the shipping preparation process and required disposal of all the individual pack cartons. When asked why the product was not bulk packed in the manufacturing center, the response was that an additional part number would have to be developed for the bulk pack and would require an additional storage location in the DC as other customers may want to order the item in individual cartons. The owners of the part number creation process ignored the fact that the DC had to create a repack location in the DC and allocate space and resources to this process. Further analysis showed that 87% of the orders for these products were in bulk quantities, not individual packs. There was no single owner of this value stream that could bring rationality to the packaging process.

Creating ownership of the value stream is critical to the success of PFEP. Ownership should sit as close to the value-added points in the value stream as possible. This is probably at the manufacturing point but could be in a distribution center, certainly not in a central headquarters located miles from both points.

Ownership also implies authority; however, this is often not the case. Pascal Dennis in his book *Getting the Right Things Done,* tells us about the Toyota A3 management process where the A3 owner often does not have authority over the process he is working on, but the A3 gives him a level of authority in the organization that is respective of the actual process owners. The same often holds true of the PFEP.

Owners of the PFEP are true owners. PFEP owners own the BOM, packaging, logistics, warehousing, and manufacturing or assembly process. PFEP owners are responsible for keeping the PFEP up to date and making sure that there is some manageable and measurable change process. PFEP owners are accountable for the efficiency, sustainability, responsiveness, and profitability of the value stream.

Creating the knowledge base of the PFEP will facilitate moving the decision-making process to the lowest levels of the organization by creating rules and a disciplined process for inventory management. The rules for inventory levels are based on customer demand not on tribal knowledge or intuition. Inventory management is no longer a "black box" process. It becomes a transparent process that is based on facts, not feelings, when you use PFEP. Inventory levels are easily explainable and manageable. Inventory reductions become a process of managing the supply chain by changing lot size or frequency, not an arbitrary budget decision; they are always process driven not emotion driven.

What Do I Need to Build What I Need?

To understand what we need, we need to approach the PFEP as if we were in a chess match. Gathering data for the PFEP is more than just gathering a part number, usage, container size, and supplier information. You need to be able to see three moves ahead of the one you are now playing in the supply chain to successfully manage your inventory.

Have you ever scheduled a job or build and gone through the setup process only to find that you are missing a component part? All the time (money) and energy spent to get that job ready to build is lost if you lack any key component. In addition, because you have scheduled precious manufacturing time and have to stop and change over to something else, you lose that time forever. Your recovery options will be limited to scheduling overtime or delaying other jobs or orders, putting your plant in a reactionary mode that continues to perpetuate itself.

The process of managing discrete orders in Material Requirements Planning (MRP) systems is further complicated once a job is delayed. Although you may believe that MRP "reserves" components for these discrete orders, MRP only dumps the components into an available pool. All components are physically available to fulfill any manufacturing order, not the discrete order for which they were originally designated to be used. The components required for the delayed job become resources for other expedited orders.

As long as job seven is delayed waiting on component C, we might as well schedule job nine, because it only needs components A and B. In the meantime, we have either lost component C or consumed component C on another job. Now we have to expedite C to replace the components consumed by the other product we produced. Meanwhile we used half of part B on another order (they were just sitting there), C arrives, and we set up again only to find now we need more of part B. In the traditional inventory environment, the worst sin you can commit is to run out of something, so you double-order B and C.

This becomes a vicious cycle. Most inventory decisions are reactions (or overreactions) to out-of-stock conditions (firefighting and expediting) that would never have occurred if they were not trying to micromanage inventories using tools that were not designed to manage manufacturing systems.

To help our clients better understand what happens when they embark on a Lean journey, we have created a Lean simulation to illustrate the process. The Lean simulation is broken into three rounds: the mass production round, the Lean round, and the PFEP round. Round one shows the problems with a mass production MRP-driven system. In round one, components are ordered for finished goods and assembled into kits or packages that are sent to the assembly floor. The simulation shows what happens when component quantities are wrong, shipments are delayed, or quality problems occur in the manufacturing process. Inevitably, most round one simulations fail to meet customer demand. In fact, in over 25 simulation rounds, no group has ever met customer demand in round one.

To succeed in meeting customer demand you have to change the way you think about a part and its components. Most facilities that we visit and work with focus on the assembly process as the major value-added activity of that organization and labor cost as the primary controllable variance The truth is that materials costs are usually 50 to 65% of the end item cost.

Takt Time

Takt time is the maximum time interval in which your customer demands that you supply a part. This time interval is always reflected in seconds, not minutes or hours. *Takt* is not a Japanese word but rather is derived from the German word *taktzeit,* which translates to "cycle time." Figure 3.2 shows the calculation for true takt time.

Within Lean, takt time is not the same as cycle time. Lean cycle time refers to the balance of work required to promote the production of a unit within the customer demand rate, or takt time. Usually, the target cycle time for a cyclical repeatable process such as a moving assembly line is 95% of the takt time. Cycle time is used to promote a structured approach to reduce disruptions that affects efficiency, quality, and value.

$$\text{Takt Time} = \frac{{}^*\text{Total Daily Operating Time}}{\begin{array}{c}\text{Required Total Daily Production For The}\\ \text{Day (Customer Demand)}\end{array}}$$

* Daily total operating time is figured on the basis of all machinery operating at 100% efficiency during regular working hours.

Figure 3.2   Toyota takt time calculation.

Takt Time Calculation Example

For this example, we say that our customer demand is 538 units a day. If we start with a shift of 8 hours (480 minutes or 28,800 seconds), we have two 10-minute breaks (600 seconds) and a 30-minute lunch break (1,800 seconds). Furthermore, our hypothetical process will use three components to create an assembly within the allocated time. We can now calculate our shift time available for production.

Shift Time Available − Breaks = Available Time (A/T) to Produce

480 − 50 = 430

430 minutes A/T × 60 Seconds = 25,800 seconds A/T per shift

We now have the two elements to calculate our takt time.

Shift Available Seconds/Customer Demand = Takt Time

25,800 Shift Available Seconds/538 Units Customer Demand
= 48 Seconds Takt Time (T/T)

Based on this calculation you have a process that makes a part from three components and the takt time is 48 seconds for that station. We understand that no process can run at a 100% efficiency rate, so we will load the process at 95% efficiency This 95% load will allow us to calculate our ideal cycle time [48 seconds × .95 (95% load) = 45 seconds] of 45 seconds. This ideal cycle time then is the load or burden we place on the operator and our line or process. This means we load 45 seconds of work content on the operator and allow her 48 seconds to complete the work building in 3 seconds of waste in every cycle. This means that if we produce 538 units per shift, we are building in 26 minutes of waste in every shift.

This now allows us to begin to think about ways to reduce the third waste in the Toyota production system: *muri*. Muri is the overburden and unreasonableness found in a process that has no standardization or concept of takt time. We reduce and eliminate muri by creating robust standard work where every process is reduced to its simplest elements for examination and further recombination through the use of Yamazumi charts.

The concept of muri probably had more to do with the success of Henry Ford's production system than the actual assembly line concept. In order for the production line to function, each station on the line had to achieve its ideal cycle time because the next station was only equipped to work on standard condition components.

Carrying this example even deeper, we stated that the operator has three components to assemble. With an ideal cycle time of 45 seconds this means the operator has 15 seconds to assemble each component. To understand the magnitude of your supply chain, you have been herding these components (your planners, buyers, suppliers, and logistics people) to get them to the operator in time for up to 16 weeks or longer.

The Role of the Supplier

You must begin to think of the supplier as an extension of the manufacturing process that is just a little farther away. Logically, if my MRP tells me I plan to build an average of 1,076 widgets per day over the next four weeks, my supplier will assume I am building 21,520 widgets (1,076 widgets × 20 working days). Unfortunately, when executing the plan it may end up looking more like the graph in Figure 3.3.

Demand in this picture averages 5,380 per week; however, actual units per week vary by two to three times. This scenario is actually very common. No one really knows what he will build next week, let alone four weeks from now. It becomes impossible to forecast or plan with any degree of confidence or reliability. Suppliers need accurate data on consumption to help you be successful. Taiichi Ohno stated that the only thing you can depend on when looking at forecasts is that they are always wrong and always change. Because we cannot accurately forecast, how do we more reliably connect end-user consumption to manufacturing through to the supply base?

This unevenness we see in the forecast is known as the second form of waste, *mura*, which is the unevenness or inconsistency found in information needed to accurately plan. Mura can be avoided by the Just-in-Time

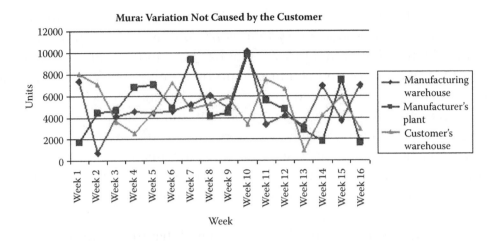

Figure 3.3 Mura.

concept of having the right part, at the right time, in the right quantity at the operator's fingertips, by utilizing a pull system to trigger the supply chain upstream of the consuming process and having strategically placed buffers as defined by the PFEP.

Your customer will not level your schedule for you; therefore, this will have to be done by your organization. To truly see the benefits of mura reduction and elimination, production leveling is just as important as smaller batch production quantities and more frequent deliveries.

The PFEP allows you to see the waste in the system and change the way you think about a component, or the *every* in Plan for Every Part. To start the PFEP, there are several questions that have to be answered about every part. Figure 3.4 give you a starting point for these questions.

Most of the information needed to populate your PFEP can be pulled from your existing MRP system. It is absolutely imperative to begin with this basic information. Often, this information does not go deep enough. You may have to take the search even further to allow the PFEP to manage your supply chain internally and externally. We typically find things such as part weight maintained by the quality department, container size and type maintained by logistics, parts used per station maintained by production engineering, and so on. Every part is going to have a story to tell; just finding all the parts of the story is the challenge here.

You have to collect as much data as you can about your parts and components in order to manage your PFEP. If you are missing weights or dimensions, or consumable components of your finished good, you will

Key Elements

Plan For Every Part

Establish a plan for every part.
There are several questions that need to be answered before you can determine this plan.

Every part has a story to tell and below are some of the questions we must ask to understand each part's story.

1. What is it?
2. How many do I use and how?
3. Are there multiple usage points?
4. What type container is it in?
5. How many do I get per container?
6. How do I order it?
7. How do I determine min/max?
8. Where does it come from?
9. How does it get to my dock?
10. How often does it arrive at my dock?
11. How do I verify the shipment?
12. What do I do with it after it arrives?
13. How do I store it?
14. How do I know how much to deliver to the usage point?
15. How do I get it to the usage point?
16. How is it presented at the usage point?
17. How do I handle empty containers?
18. Where do I store them?
19. Do I have too many or not enough containers in circulation?
20. When do I ship empties back to the supplier?
21. How does replacement order work?
22. What do I do with overflow of parts?
23. What if I have a shortage?
24. Who is involved in this process?
25. What if part is damaged who do I call to determine disposition?
26. How do I get replacements?

These questions must to be answered before a material handling flow can be established for all parts.

Master Worksheet

Part Number

Usage

Qty. Container

Order Timing?
Daily
Weekly
Monthly

Order Method?
Pullcard
Usage
Forecast
EDI
Guess

Investigation (Problem Solving)?
Shortage
Rundown
Other
Overflow
Reduction plan
Other

Supplier: My Part
Location
Frequency of delivery
Drive time
% Dedicated to my part
Transportation type

Logistics?
Dedicated
Milkrun
Contract
Other
Does each truck have contact method?
Cell phone
Pager
GPS
Other

Shipment Verification?
ASN
Manifest
Bill of lading

Usage Delivery Method?
Small (card)
Call
Sequenced

Storage Layout?
Lane
Rack
Direct delivery

Dock Management?
Schedule unload/load out
5S – visual factory
Daily walks
Safety
Problem solve

Figure 3.4 Plan for every part questionnaire.

not be able to manage your entire value stream; you will be managing pieces of the value stream, reacting or firefighting the components you are not managing.

What Do Suppliers Need?

Suppliers (and even customers) want the same thing you want: predictability and stability. If you forecast 1,076 widgets per day for the next four weeks but order 3,000 one day, nothing the next two days, and 2,380 on day four you have technically met your forecast. However, you have also disrupted your supplier's manufacturing process by creating lumpy demand and not smoothing the flow of material and allowing them to smooth their manufacturing process. Simply stating, "I am the customer so they have to meet my needs" is not the solution. The solution is to work with *your* customer to level the demand. This introduces stability and leads to predictability. You also need to work on your manufacturing process so that you do not have to run a large batch size of 5,380 once per week, instead running 1,076 per day.

Understanding the Bill of Material to Populate the PFEP

There are two basic structures to the BOM. The simplest form is flat with few levels, as shown in Figure 3.5. This is the easiest way to track inventory for the engineering staff and this simple structure helps in meeting financial goals more easily. The second is deep so that we can see all the engineering levels involved in the process. If you look at the Toyota design structure, you will see design group, index, and variation at levels one, two, three, and so on. Each part on the vehicle fits within a functional group.

For example; Figure 3.6 shows that you would find a wiring harness in design group 8200, and within that group depending on the area of the vehicle, say the cowl wire, you would find it in index 11. This translates to the first five digits of the 12-digit part number 82110-. Figures 3.6a and 3.6b are two examples showing the variation portion of the group and index used to break out the many cowl wires, usually by destination or option content.

If you were talking about a 49-state Camry cowl, the next five digits would be -06010, Avalon-07010, and so on, based on the particular vehicle on which it is used. The part number also allows you to tell the design center involved in the design of the part: if it was designed in Japan it would

Part Number	Description	Level
LG1000	Floor Post	1
LG1001	16 Peg Floor Panel	1
LG1002	12 Peg Black Block	1
LG1003	Angled Yellow Block	1
LG1004	RH Cowl Panel	1
LG1005	LH Cowl Panel	1
LG3001	Axle	1
LG3002	Tire & Wheel Sub-Assembly	1
LG1320	Angled Corner Post	1
LG1323	Front Grill	1
LG1625	Left Door Panel	1
LG1625	Right Door Panel	1
LG4200	Rear Bumper Post	1
LG1325	6 Peg Black Block	1
LG1627	8 Peg Block Block	1
LG1330	Exhaust Port	1
LG1333	2 Peg Red Block	1
LG1332	3 Peg Corner Block	1
LG1327	Front Fender	1
LG1324	2 Peg Black Block	1
LG1331	Tail Light	1
LG1329	Rear Deck	1
LG4201	Rear License Plate Frame	1
LG1322	Hood Intake	1
LG5000	Roof Piece	1
LG1326	Spoiler	1
LG1328	8 Peg Rear Deck	1

Figure 3.5   Typical bill of materials.

Group	Index	Variation	Part Number
8200	11	10	82110-06010-00
	11	20	82110-06020-00
	11	30	82110-06030-00

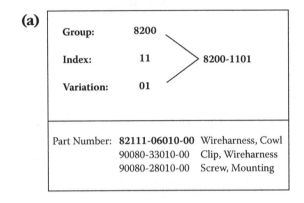

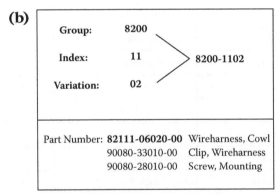

Figure 3.6 Toyota group and index model.

be the -06010, if it was designed in North America it would be -AA010. So the part number for a 49-state cowl wire would be 82110-06010-00. The last two (-00) digits are left for color code or Japanese supplier codes. These part numbers are not only used to designate an identifier for the part, they show a relationship to the entire structure of the car. This is an intelligent design format.

The Toyota design structure is far more advanced than anything we have seen in any other business. The way a part number is structured and designed is part of the visual controls Toyota builds into every process. Toyota's BOM structure makes it very easy for decisions to be made at a low level by recognizing how and where the part is used in the assembly process.

Often, outside groups will unintentionally have an adverse influence on the design of your part structure. Engineers will often use phantom part numbers for subassemblies to hide variation. Finance will often recommend expensing hardware or packing supplies, believing that it simplifies materials management. We cannot emphasize enough that PFEP means *every* part. The BOM needs to be as flat as possible, removing and simplifying layers allowing all components to be visible.

Packaging, whether expendable or returnable, should be managed just like any other inventory item. Cardboard is a significant expense for most manufacturers, not just our environment. One industrial distributor we worked with reduced cardboard expense by $500,000 per year, just by removing variation in box size and managing inventories by using a PFEP. Instead of dozens of pallets of unused boxes stacked in the warehouse, the PFEP told them how many of each size they needed, based on their historical consumption patterns. The distribution center was able to sell back almost $40,000 in obsolete (to them) box sizes to their supplier.

More often than not, we find fasteners are either VMI or simply expensed without thought as to how to use the BOM to ensure a quality finished good. Making fasteners part of the BOM ensures that the correct part is identified in the correct quantity for the process and product that is being assembled. This facilitates creating standard work for the operator and material handler. Knowing the right quantity needed and consumption rates will allow you to create a timed delivery sequence and pattern for all parts, including fasteners, packing, and labeling materials. A common misunderstanding of VMI is that once you turn it over to an outside firm to manage for you, you don't have to do anything. VMI still requires some audits, among other things.

Once you identify your BOM and assign an appropriate level, the next step is to understand where the part is best used in the value stream. This is accomplished through your value stream map, assigning loops to understand the flow of material. Inventory loops run from supermarket to supermarket upstream through the value stream starting with the distribution center/finished goods to the customer, distribution center/finished goods to the point-of-use flow rack at the assembly process, and so on.

The inventory loop from the distribution center to the customer should be filled with only what the customer orders or ordered. Ship the customers what they want, when they want it, and in the quantity they request. In the next chapter we begin to dissect the inventory loops starting from the manufacturing plant to the distribution center/finished goods supermarket. We also begin to break down the items and variables required to make the PFEP user friendly, making it easy to understand every part in your system in much greater detail than you understand today.

Chapter 4

Managing Loops

To understand what component is needed and where it is used, we have to build a representation of what the flow looks like from the customer all the way back through the supply chain to the raw material supplier (see Figure 4.1). To see the whole picture, you first have to draw, or map, the value stream. The current state value stream map will help you understand your process families, constraints, capacity, and capability to support the end customer. By mapping the current state, you will learn what you are capable of producing by focusing on the door-to-door flow of material and information while breaking down each process capability. You will also identify where your bottlenecks are in the flow.

Once we have the current state map drawn and data blocks filled in, we can begin to populate the Plan for Every Part (PFEP) by determining the finished goods popularity code, design level, and model life cycle codes and usage across the supply chain. Matching your PFEP with each set in the value stream map will allow you to push the PFEP far beyond just managing finished goods and their components: if you build the PFEP correctly, understanding its many parts, you can plan, schedule, and manage your supply chain more.

Value Stream Mapping

Your value stream is your roadmap to where you will be going on your Lean journey. Few of us begin a family trip by jumping in our car and driving until we run out of gas; we normally have a destination in mind. To get to that destination (our future state map) we first need to know where

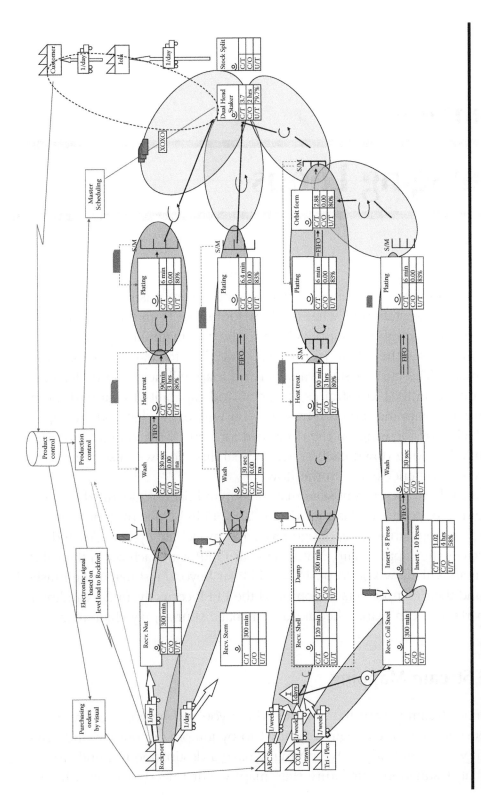

Figure 4.1 Managing the loop.

we started (our current state). Trying to create a route map without a clear understanding of where we started and where we want to end will lead to chaos and failure. The value stream map is all the value-added and non-value-added actions required to get a product through the flow of your facility from the raw material supplier to the end customer. You map both the material flows as well as the information flows required to support production of your product. The goal in your future state map is to remove the waste in man, machine, and material (and ultimately scheduling to only one process in the flow: your pacemaker process), adding velocity to your supply chain.

If you are not comfortable with value stream mapping, we recommend that you read *Learning to See* (Rother and Shook, 1999) and draw four or five current state maps before you continue. The value stream map is your roadmap to identifying the non-value-added steps in your process and is an integral part of creating your PFEP. Trying to create a PFEP without understanding your value stream will be ineffective. Begin by looking at the end of the value stream, the point closest to your final customer and the single finished part at the end of the process.

Once you have identified your final customer, move back upstream in the flow from there; some organizations maintain a finished goods supermarket within their four walls, whereas others produce and ship directly to a Distribution Center (DC) where the parts are maintained awaiting shipment. Often this DC acts as a finished goods supermarket for multiple manufacturing sites in order to speed the shipment of combined items and to form one purchase order to one end customer under one shipment.

If you build and ship directly to a DC with no finished goods stored at your facility, the DC is in fact your finished goods supermarket. The next process in your value stream (the DC) needs to trigger your production signal. If you have a finished goods market between you and the next customer, then the inventory level at that market will signal the replenishment. The inventory trigger points are determined by factors such as delivery frequency, delivery cycle time, order lots, and safety stock, the logistics side of your inventory cycle.

There are some Lean practitioners who say you should never use your DC as the pull on your manufacturing process. You must be careful with generalizations. Even Toyota uses their dealer inventories as a form of pull, not to replenish a specific vehicle that was sold to a customer, but the total inventory becomes a trigger to determine production rates. Toyota's goal is to keep a specific number of days' inventory in the field; it is up to the dealers to identify the specific make, model, and option configuration they want to stock. Toyota uses this information to speed up or slow down the production line, not to decide what to build.

For the rest of us, we need something better. When you cannot flow product into the market, you must have a method to pull product into the market. For that reason, companies must develop a robust pull system to keep product on the shelf. If your products are highly seasonal you will need to build to a supermarket (depending on capacity, this could be a long lead time) to complete your buffer and then allow the season to ramp down as you change over to the next seasonal item. In either case, your supermarket, whether it is internal or external, must be one of the trigger loops you consider.

For the purpose of your PFEP it does not matter if the demand is coming from a DC to replenish or straight from the customer to be pulled from a finished goods market; the finished good product that is shipped to the customer (whatever the definition of the customer is) is the level 1 part on your PFEP, as seen in Figure 4.2.

The components that are pulled to make the finished part are your level 2 parts, and the materials/components used to make the level 2 will be your level 3, and so on. You will need to determine if your component parts are unique or common to a product family. Product families, as described in the book *Learning to See*, are a group of products that flow downstream through similar processes and across common equipment. To keep your system simple use "U" for *unique* and "C" for *common*, and then add this information to your PFEP, as shown in Figure 4.3. When we refer to unique, we are referring to component parts that are unique to a finished part number or a line and are only used in the production of a unique finished part number. A common part is one that is used across multiple models or parts.

In Figure 4.4, the column titled "Make or Buy" indicates if the source of the part is internal (make) or externally sourced (buy). If you are purchasing and receiving components from a sister plant from within the same parent company, this is considered an external source. We use "M" for *make* and "P" for *purchased*. The PFEP at this point can start to tell you how to lay out primary and secondary machines or processes with changeover and production wheels for the replenishment of needed components without overproducing.

When you know the demand for a finished part number, you can then calculate the demand for the components. You can now start to calculate the desired level of component inventory in the producing department and understand your delivery intervals for replenishment.

Component Part Number	Component Description	Finished Part Number	Make or Buy	Design Level	Finished Product Description	Life Cycle Codes	Unique or Common
3002421	SPRING VALLEY ARBOR	7002421	M	1	SPRING VALLEY ARBOR	0	U
4103252	4×8 SMALL PAD	7002421	P	2	SPRING VALLEY ARBOR	0	U
1103300	3/4×3/4×10 TOE BOARDS	7002421	P	2	SPRING VALLEY ARBOR	0	U
1103446	PALLET BOARD 5/8 × 6 × 50	7002421	P	2	SPRING VALLEY ARBOR	0	C
4103278	BOX CROSS BEAM SEVEN	7002421	P	2	SPRING VALLEY ARBOR	0	C
1102968	1.375×1.375×22.5 EXTR .065 WALL W/LT	7002421	M	2	SPRING VALLEY ARBOR	0	U
1103005	1.375×1.375×22.5 EXTR .065 WALL WIP	1102968	M	3	1.375×1.375×22.5 EXTR .065 WALL W/LT	2	U
1103006	1.375×1.375×22.5 EXTR .065 WALL WIP	1102968	M	4	1.375×1.375×22.5 EXTR .065 WALL W/LT	1	U
1103445	UPRIGHT INSERT 5/8×6×46 WOOD	7002421	P	2	SPRING VALLEY ARBOR	0	U
4103681	BAG 17 SHRINK 154×55×.0045 WHT OPAQUE	7002421	P	2	SPRING VALLEY ARBOR	0	C
4102575	POLYROLL 60 WHITE OPAQUE NON-LIP UVI	7002421	P	2	SPRING VALLEY ARBOR	0	C
4101972	LABEL "STRAP HERE"	7002421	P	2	SPRING VALLEY ARBOR	0	U
4101549	BOX UP RIGHT 15	7002421	P	2	SPRING VALLEY ARBOR	0	U

Figure 4.2 PFEP component part numbers.

Component Part Number	Component Description	Finished Part Number	Make or Buy	Design Level	Finished Product Description	Life Cycle Codes	Unique or Common
3002421	SPRING VALLEY ARBOR	7002421	M	1	SPRING VALLEY ARBOR	0	U
4103252	4×8 SMALL PAD	7002421	P	2	SPRING VALLEY ARBOR	0	U
1103300	3/4×3/4×10 TOE BOARDS	7002421	P	2	SPRING VALLEY ARBOR	0	U
1103446	PALLET BOARD 5/8 × 6 × 50	7002421	P	2	SPRING VALLEY ARBOR	0	C
4103278	BOX CROSS BEAM SEVEN	7002421	P	2	SPRING VALLEY ARBOR	0	C
1102968	1.375×1.375×22.5 EXTR .065 WALL W/LT	7002421	M	2	SPRING VALLEY ARBOR	0	U
1103005	1.375×1.375×22.5 EXTR .065 WALL WIP	1102968	M	3	1.375×1.375×22.5 EXTR .065 WALL W/LT	2	U
1103006	1.375×1.375×22.5 EXTR .065 WALL WIP	1102968	M	4	1.375×1.375×22.5 EXTR .065 WALL W/LT	1	U
1103445	UPRIGHT INSERT 5/8×6×46 WOOD	7002421	P	2	SPRING VALLEY ARBOR	0	U
4103681	BAG 17 SHRINK 154×55×.0045 WHT OPAQUE	7002421	P	2	SPRING VALLEY ARBOR	0	C
4102575	POLYROLL 60 WHITE OPAQUE NON-LIP UVI	7002421	P	2	SPRING VALLEY ARBOR	0	C
4101972	LABEL "STRAP HERE"	7002421	P	2	SPRING VALLEY ARBOR	0	U
4101549	BOX UP RIGHT 15	7002421	P	2	SPRING VALLEY ARBOR	0	U

Figure 4.3 **Unique and common PFEP components.**

Component Part Number	Component Description	Finished Part Number	Make or Buy	Design Level	Finished Product Description	Life Cycle Codes	Unique or Common
3002421	SPRING VALLEY ARBOR	7002421	M	1	SPRING VALLEY ARBOR	0	U
4103252	4×8 SMALL PAD	7002421	P	2	SPRING VALLEY ARBOR	0	U
1103300	3/4×3/4×10 TOE BOARDS	7002421	P	2	SPRING VALLEY ARBOR	0	U
1103446	PALLET BOARD 5/8 × 6 × 50	7002421	P	2	SPRING VALLEY ARBOR	0	C
4103278	BOX CROSS BEAM SEVEN	7002421	P	2	SPRING VALLEY ARBOR	0	C
1102968	1.375×1.375×22.5 EXTR .065 WALL W/LT	7002421	M	2	SPRING VALLEY ARBOR	0	U
1103005	1.375×1.375×22.5 EXTR .065 WALL WIP	1102968	M	3	1.375×1.375×22.5 EXTR .065 WALL W/LT	2	U
1103006	1.375×1.375×22.5 EXTR .065 WALL WIP	1102968	M	4	1.375×1.375×22.5 EXTR .065 WALL W/LT	1	U
1103445	UPRIGHT INSERT 5/8×6×46 WOOD	7002421	P	2	SPRING VALLEY ARBOR	0	U
4103681	BAG 17 SHRINK 154×55×.0045 WHT OPAQUE	7002421	P	2	SPRING VALLEY ARBOR	0	C
4102575	POLYROLL 60 WHITE OPAQUE NON-LIP UVI	7002421	P	2	SPRING VALLEY ARBOR	0	C
4101972	LABEL "STRAP HERE"	7002421	P	2	SPRING VALLEY ARBOR	0	U
4101549	BOX UP RIGHT 15	7002421	P	2	SPRING VALLEY ARBOR	0	U

Figure 4.4   Make or buy PFEP components.

ABCs of the Part Number

Now you need to assign popularity codes for the inventory. We are talking about the 80% versus 20% rule, meaning that typically 80% of your business is 20% of your finished part numbers. This means we need some way to break down the finished goods part numbers, and their components, into a matrix that can be reviewed and studied easily to help make sound rational decisions about your inventory. For this we use a popularity coding system of A, B, C, and D. A part is coded an "A" item when it is a fast mover a with a predictable level of demand. A part is coded a "C" item when it is a slow mover, meaning it is a part that sells infrequently. The "B" parts are the hardest to categorize because some fall close to an "A" and some close to a "C." "D" parts are usually stocked parts that move infrequently but are considered essential to the catalog offered to customers. Figure 4.5 shows the typical hierarchy.

Start your classification of part numbers closest to the finished customer or with your finished goods, because customer demand is what sets your takt time. Finished goods must be managed at the SKU level. Often, however, we find that A, B, C, and D classifications are insufficient to explain inventory variability. Within the finished goods, you may see this variability requiring you to break the overall categories of A, B, C, and D down into smaller, more manageable categories such as A1 and A2, to help with flow throughout the supply chain. We typically set the A1 category as the top 3 to 5% of sales and categorize each division as the next 5% until we get to the 50–55% of part numbers. The bottom tier (D2) normally accounts for less than 5% of sales. Figure 4.6 illustrates a typical part number distribution with a total of 628 finished good part numbers, with A1s as the fastest movers, and a total of 11,782 finished good part numbers, and D2s as the slowest movers. As you can see in Figure 4.6, 53% of the part numbers represent

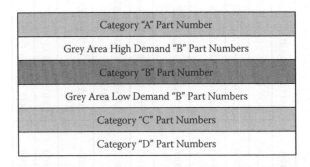

Figure 4.5 Popularity codes.

Popularity	Days on Hand	Part Number	% of Part Number
A1	25.71	628	2.8
A2	27.21	1,231	5.6
B1	29.86	1,230	5.6
B2	32.99	2,060	9.3
C1	38.67	1,088	4.9
C2	43.54	2	6.9
D1	54.06	2,533	11.5
D2	183.28	11,782	53.4
Total	54.415	22,052	100.0%

Figure 4.6   Life cycle codes of PFEP components.

the largest block of inventory on hand, often making up over 50% of the inventory dollars a company has invested yet these parts often represent less than 5% of sales.

There are rules to establishing an A, B, C, and D classification matrix, and by examining the pattern of a part's sales history and forecasted sales trends, you will be able to decide where to cut off the A's and start the B's. Although subjective, it is a very important planning decision. This decision has the biggest impact on the finished goods market inventory, required space, manpower, and budgeting as well as the way you produce product and buy components for the A and B items. Your decision of where you set the cut-off volume for your A and B part numbers will directly affect the total dollars of inventory your organization holds, essentially setting your inventory plan. The faster the inventory turns on these fast movers, the more cash you have for other aspects of your business. For most C items and all D items, we recommend that you procure components using the typical Materials Requirements Planning (MRP) method and schedule these items in for production using the "white space" or available time recognized by using the PFEP, changeover, and production wheels.

You should consider holding a larger buffer of the middle to lower B items in your finished goods market and fewer of the upper B and A items to allow for faster turnaround on the production wheel while maintaining the quality and reducing the obsolescence of high runner parts. This does not mean to load the finished goods market up with all your B items to keep from changing over your machines. Rather, use the PFEP to help plan and maintain a constant inventory turn of all products.

In simpler terms: The A's are steady and regular volume, say daily. B's are not as frequent and not necessarily high volume, but they do have a regular pattern and therefore you would want to stock them in a supermarket. C's are unpredictable and would be built to order.

The coding becomes more difficult when you start working on your design level 2 and level 3 parts. It is critical that you break down the level 2 parts into "make" and "purchased" parts. The make parts now refer to those items *you* have to produce, and usually in a batch method, to support final assembly. We cover this more in our section on Dynamic Capacity Planning (DCP) in Chapter 8.

For purchased components, let's work through a scenario. You have a C finished part number, sell one a year, consisting of 50 components, and what makes the finished part number a C is one single component, which means that the other 49 components are used on all the other models you make, and therefore those components have to be considered even for the C item we are manually scheduling into the assembly system. If we have decided as an organization not to hold any C items, either finished or components, we have to order these items in the amount required to build the one unit.

Now, if your sales team has priced the final unit, you are now obligated to sell that unit for the quoted price. What if you go to buy that one component, just one of them, and the supplier quotes you a price 10 times the price of buying 100 of them? Two things can happen here; first, you buy just the one component and lose money on the sale, then bring sales into the loop and set guidelines for pricing C finished part numbers; second, you buy the 100 and 100 years' worth of components to manage and sit around costing you every time you cycle count it, move it, inventory it, and so on. We know of no finished part number that has a life cycle of 100 years.

Life Cycle Code

You must have a method for managing a product's life cycle—not only components but finished goods as well. Life cycle management tells us that when a product reaches its logical end, it must be discontinued. Products should not be "immortal." There comes a point when it is no longer profitable to carry inventory on every item it is possible to manufacture. Your customer may want that product to repair his 1968 Ford, but he really does not expect it to be a stocked item. He will be willing to wait for it if you give him a realistic promise and then deliver it at the promised time. Furthermore, when components are no longer common, there is a point at which it is not economical to continue to support the manufacture of these items. As difficult and painful as it might be, sometimes products must be allowed to be discontinued.

Toyota uses a Life Cycle Code (LCC) to manage changes to parts or materials with extreme accuracy and minimal obsolescence. A life cycle code of "1" was assigned in the BOM for new part numbers, a "2" for build-out part numbers, and a "0" for carryover part numbers. This function of adding the codes to the PFEP will need to be developed for your facility. At Toyota, it was the design engineer's responsibility to add the LCC 1 and part numbers to the system, code the buildout part as a 2, and issue the Engineering Change Instruction (ECI) from the old to the new part. Production control would receive the ECI and review the changes, contact the supplier, and begin the coordination of ramping up the delivery of the new part and the ramping down of the buildout part.

In our example, Figure 4.7, we have used the Toyota coding system. You will want to align the LCC code in the PFEP close to the supplier section. This will help when trying to determine the effect an add 1 or delete 2 part number has on the supply chain, by seeing it next to the supplier location; to help plan delivery cycles, frequency, and time; and to help reduce inventory and floor space. For internal suppliers, use the machine or department name as the supplier name. Order frequency for internal suppliers will be the Every Part Every Interval (EPEI).

The Right Quantity—Daily Usage Rate

The final piece of the PFEP is identifying usage. Usage is at best a guess. It can be based on historical consumption, but this can be misleading. If you use average historical consumption you will miss seasonal or promotional trends. If you use forecast from your customers, your plan will only be as good as your customer's ability to look into the future. Toyota used a fixed planning horizon so they could provide a detailed firm forecast to their supply base. This fixed horizon went as far as 13 weeks, with four "frozen" weeks. In recent years even this has changed as Toyota has tried to reduce cycle times and increase manufacturing flexibility. In reality there was very

Component Part Number	Component Description	Finished Part Number	Make or Buy	Design Level	Finished Product Description	Life Cycle Codes	Unique or Common
3002421	SPRING VALLEY ARBOR	7002421	M	1	SPRING VALLEY ARBOR	0	U
4103252	4×8 SMALL PAD	7002421	P	2	SPRING VALLEY ARBOR	0	U
1103300	3/4x3/4x10 TOE BOARDS	7002421	P	2	SPRING VALLEY ARBOR	0	U
1103446	PALLET BOARD 5/8 × 6 × 50	7002421	P	2	SPRING VALLEY ARBOR	0	C
4103278	BOX CROSS BEAM SEVEN	7002421	P	2	SPRING VALLEY ARBOR	0	C
1102968	1.375×1.375×22.5 EXTR .065 WALL W/ LT	7002421	M	2	SPRING VALLEY ARBOR	0	U
1103005	1.375×1.375×22.5 EXTR .065 WALL WIP	1102968	M	3	1.375×1.375×22.5 EXTR .065 WALL W/LT	2	U
1103006	1.375×1.375×22.5 EXTR .065 WALL WIP	1102968	M	4	1.375×1.375×22.5 EXTR .065 WALL W/LT	1	U
1103445	UPRIGHT INSERT 5/8×6×46 WOOD	7002421	P	2	SPRING VALLEY ARBOR	0	U
4103681	BAG 17 SHRINK 154×55×.0045 WHT OPAQUE	7002421	P	2	SPRING VALLEY ARBOR	0	C
4102575	POLYROLL 60 WHITE OPAQUE NON-LIP UVI	7002421	P	2	SPRING VALLEY ARBOR	0	C
4101972	LABEL "STRAP HERE"	7002421	P	2	SPRING VALLEY ARBOR	0	U
4101549	BOX UP RIGHT 15	7002421	P	2	SPRING VALLEY ARBOR	0	U

Figure 4.7 Life cycle codes of PFEP components.

little fluctuation within this 13-week horizon (the recession that started in 2008 notwithstanding).

Most of us do not have the luxury of an accurate frozen forecast. So we manage inventory in this environment by using an inventory buffer. PFEP is the tool that helps us establish this buffer. Most elements of the PFEP are "fixed"; that is, they are known and predictable. Transit times, delivery frequency, and reliability (quality) are all easily calculated. It is the buffer or safety portion that allows us to manage the inevitable fluctuations in demand. As you remember from earlier, Toyota calls this *mura* and it is defined as variation not caused by customer demand. The amount of this buffer is your decision alone; there is no standard formula. At Toyota this buffer was set as low as 27 minutes on the assembly line. They could do this because they control the logistics pace and frequency, both internally and externally, and because production schedules, once set, were never changed. For the rest of the manufacturing world (us), our definition for buffers is "the amount of time it will take to re-plan the previous process or expedite parts through the manufacturing and logistics networks." This is expressed as days, hours, or minutes, depending on the length of the expedited fulfillment cycle.

Even promotions and seasonality are predictable. They can be managed by changing the size of the buffer leading into the promotion and reducing buffer size as seasons wind down. Remember, buffer stocks are based on the average Daily Going Rate (DGR). PFEP assumes replenishing based on what was actually consumed by the downstream process. Ideally you would want to adjust DGRs based on known factors. However, unless you have a robust system that allows you to easily make changes to DGRs, up or down, it is often easier to manage these fluctuations by changing buffer sizes.

Why Do I Need All This Stuff?

Most off-the-shelf MRP systems have nowhere to build a PFEP-style data retrieval system. They were not designed for this function. MRP is needed by the supply base as it is a great forecasting tool; however, MRP is not a scheduling tool.

When you consider the current flow of information needed to make rational decisions on a part number in your value stream, you will begin to see the waste in the process of having to track down this information. PFEP data are contained in multiple places scattered around different departments and among people with tribal knowledge of the day-to-day running of your value stream and has many owners. These owners do not know how their independent actions affect other areas of the business.

Failing to see the entire picture often leads to decisions that cost the company money. As we said earlier, typical business models have silos of knowledge strung among them. The old paradigm, "I don't care how you get it done, just get it done," plays heavily in the way we do business today. What we want to drive is standardization in the decision-making process, placing decision making at the lowest level of the organization. This is only possible when all the information is present to lead to a successful decision. We always recommend that an organization start their PFEP in a Microsoft Excel worksheet and then as it grows larger, move it into a Microsoft Access database format. Both of these options allow for the data to be secured and manipulated to meet the demands of the process. MRP can only be rendered this useful if you have a very dedicated programmer and usually at a large cost.

When you begin to look at the data that are needed for a PFEP, go back to the future state map and look at the material loops established for the material flow. This will drive questions about what information is needed for that loop in the value stream.

- How do the actions of the sales team play into my value stream plan?
- Who is setting the goals of the organization?
- Are we a private or public company?
- What are the financial goals for inventory?
- How is the performance of the inventory team measured?

These questions play a dramatic role in the way you set up the PFEP parameters. Traditional metrics of the business, including labor and overhead absorption, will usually not align with the PFEP. Focusing on labor or material absorption will cause you to make bad decisions in material planning. Achieving a one-cent-per-piece savings on a component or buying from an offshore source leads to long supply chains, potential damage from moving material several times, storage costs, and quality issues. Finding defective components in a 16-week-long supply chain will cause any savings to be absorbed in the cost of expediting replacement components. Long lead times mean more inventory and more inventory leads to inefficiency and waste.

PFEP will force you to look at every part number on your bill of material. You cannot take the shortcuts of expensing items such as hardware and shipping containers. These items are the pennies in the Lean metrics. Although the framework of the PFEP is generic, it has to be customized to fit each business model that uses it. This requires you to think deeply of your bill of materials to the lowest-level component. This thinking also requires you to understand and begin to manage all of the processes in your inventory loop, including inventory levels, logistics, and material handling.

We believe there are standard pieces of information that must be on every PFEP for a company and plants within that company. As we try to point out, there are then the added data points that are important to a particular business and that is the part that makes the PFEP unique. One reason this is important: companies have not realized the opportunities that exist when the standard elements of a PFEP could be seen together for a company of, let's say, 10 plants. Opportunities in logistics, suppliers, and so on become very obvious.

Information that affects the value stream comes from many different sources, but all of the information affects the value stream. The book *Learning to See* points you to the pacemaker process as the scheduling key, building your PFEP around it. We want you to move even further with this thinking by using the individual inventory loops within your value stream (supply chain) when you are planning the supply chain.

As you can see from the items and variables we have covered, there is a lot to managing every part through every loop you have in your value and supply chains. It is important to know your inventory loops in order to structure your PFEP in a methodical order that is easy to understand and at the same time covers all the variables you will encounter. The structure of the PFEP depends on accurate information to enable you to make quick, concise decisions about your business. The decisions are what will allow you to become flexible in your market and help ensure your success in a Lean implementation.

Chapter 5

Finished Goods Planning

Finished goods planning is the most complex planning area of the Plan for Every Part (PFEP). This requires us to deeply understand our manufacturing capability so that we can define our current manufacturing frequency and explore ways to increase this frequency without adding cost to the supply chain. That does not mean it should be skipped in the planning process; in fact, it is the one critical element that will pull demand through the entire supply chain. It is not an understatement to say that unless you can stabilize your finished goods planning process you will never successfully implement a pull system. To get started you will need the following key pieces of data.

Manufacturing Planning Time

Manufacturing planning time is the time it takes you to administratively create and release a manufacturing schedule to the shop floor and includes your "normal" planning interval. This is known as your *planning horizon*. Planning horizons are nothing more than how much advance notice a process requires so that machines and resources can be scheduled in the sequence that maximizes output. If you are extremely flexible, with little changeover time, this might be hours or days. Most organizations complete their schedules in weekly buckets (weekly for a five-day operation = five days), meaning we would never run a discrete item more than one time every five days.

Manufacturing Frequency

This is the *interval* in *every part every interval*. Ideally you want to run every part every day. This is possible when you have only a few finished part numbers or your finished goods are combinations of common components. If your business is driven by variation, you will want to use A, B, C classifications and determine your run frequency. For a high volume A item that is scheduled to run weekly, the frequency would be five days.

Transportation Time

If you manage your logistics, using window times for pickup and delivery, then this is a finite definable number. If you use Less than Truck Load shipments (LTL or common carriers), there is a wide variance. If the variance is predictable, with only the occasional failure, use the most consistently repeatable number for transit time and then buffer for the variation until you can remove this waste from your transportation network. In other words, if the lowest repeatable transit time from North Carolina to Arkansas is two days, but can fluctuate to as many as six days, we use two days as the transit time, but will need to add as many as four additional days to our buffer stock (safety). Also consider shipping frequency as well as transit time. If your logistics are unreliable, find another carrier that you can rely on (Distribution Center [DC] located 200 miles away, one shipment per week, Transit Time = one day).

Put-Away Time

Often overlooked, you need to consider the time it takes to receive, sort, and put products away in your DC. This is often more than a few hours and depending on how many people you can put into the process, the size of your DC, and the complexity of the shipment, can run into multiple days. In our example we use a half day.

Buffer (Safety)

We know that our planning horizon is five days and our manufacturing frequency is weekly. Because transit times are minimal in our example, we use 10 days as our safety to cover order variation.

Supply Chain Cycle Time

Therefore, our material flow loops in this example are manufacturing planning time of 5 days, manufacturing frequency of 5 days, transportation time of one day, put-away time of a half day, and buffer or safety of 10 days for a total cycle time of 21.5. Each loop is independent and should be addressed separately as part of the overall supply chain value stream. In other words, you should not independently reduce manufacturing frequency without understanding the impact of this action on your transportation or distribution center.

- Manufacturing Planning Time   5 days
- Manufacturing Frequency   5 days
- Transportation Time   1 day
- Put-Away Time   0.5 day
- Buffer (Safety)   10 days
- Total inventory in the cycle: 21.5 days

This example includes all of the factors you need to consider in finished goods inventory planning. From here you need to establish a kanban plan for managing the inventory. The kanban is the tool you will use to manage your replenishment signals. Kanban can take many forms, but it represents the unit of measure for planning purposes. The kanban can represent a standard pack, a standard run quantity, a card, or a container or rack. The key is that it has to represent the same unit for a part or component throughout the loop.

In our example, let's assume our average daily sales rate is 2,500 pieces and our standard pack is 250 pieces to a box. Customers can order any quantity from one to any number and no incentive is given for standard pack purchases. A kanban equals one standard pack of 250 units. Using the information from the above example, the inventory represented for each step of the supply chain is as follows:

- Manufacturing Planning Time: 50 kanban
- Manufacturing Frequency: 50 kanban
- Transportation Time: 10 kanban
- Put-Away Time: 5 kanban
- Buffer: 100 kanban.

Inasmuch as each kanban represents 250 pieces, our total inventory in this supply chain is 53,750 pieces.

Based on this, the standard manufacturing run should be in the range of 12,500 pieces. (250 pieces × 50—Manufacturing Frequency—If our frequency is weekly we would run approximately 12,500 every week, every interval). This is what should drive the economic batch size. Perhaps we have a rule that says we cannot run this unless we have demand for 15,000 units. We can either change the manufacturing frequency to six days, or we can allow the buffer to absorb the variation. At any rate, we want to build based on kanban consumption, that is, what was shipped to the customer. We can also use these data to set up the warehouse locations. Using standard container sizes for the finished goods, we can calculate the pick-size footprint and size appropriately. Knowing we need to keep 100 containers as the buffer, and with a weekly production batch of approximately 50 boxes, it now easy to create standard footprints to reduce replenishment or restocking times in your warehouse.

As you can see in Figure 5.1, planning the finished goods loop automatically overlaps the previous loop, leading you to understand the rules and limitations

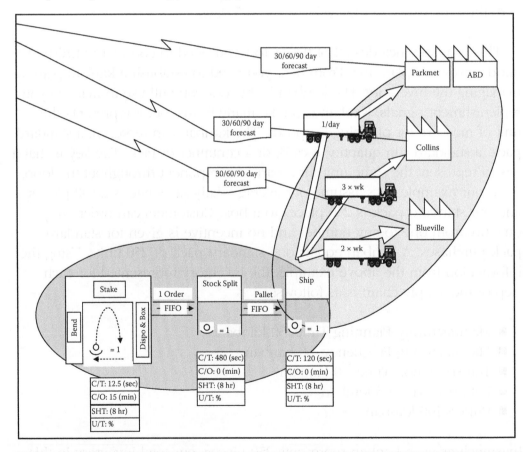

Figure 5.1   Finished goods loop.

of the preceding process. You cannot approach this from the receiving doors; you have to start from the shipping doors. As the rules of the inventory loop become apparent the question of manufacturing batch sizes becomes easier to answer. These finished goods inventory rules are similar, no matter what your distribution channel looks like. Every segment of the channel has an associated inventory that must be considered when doing your inventory planning.

Not only does transit time affect your loop, but shipping frequency has upstream and downstream implications as well. If you only ship to a DC weekly, why would you plan to build a part daily? In this case, logistics will dictate a once-per-week manufacturing cycle. If you try to build daily, you will only build inventory on your dock. Failing to consolidate these daily runs will increase handling at your DC, creating unnecessary, non-value-added handling of the product. On the other hand, if you ship daily, you will want to build daily. This will allow you to reduce not only the manufacturing frequency (from five days to one day) but your buffers as well (you will no longer have to wait up to five days for the product to be built again). Applying these ideas to the above example, changing to daily shipments and running the product in smaller batches daily will result in an inventory profile that looks like this:

- Manufacturing Planning Time: 50 kanban
- Manufacturing Frequency: 10 kanban
- Transportation Time: 10 kanban
- Put-Away Time: 5 kanban
- Buffer: 50 kanban

Inasmuch as each kanban represents 250 pieces, our total inventory in this supply chain is 31,250 pieces or a reduction in total inventory of 22,750 pieces.

Now we hope you are beginning to see why arbitrary inventory reduction targets based on day's coverage or financial value are irrelevant. On the finished goods side, if you want to drive inventory reductions, you can only do this by reducing one of the components that make up the cycle. You can only reduce buffers (without potentially affecting customers) if you can reduce manufacturing or transportation cycles. Reducing any other piece of the cycle will reduce inventories. At some point you will arrive at an optimum inventory position, and at that point the only variation will be consumption or demand rates. If demand goes up, your daily going rate increases and your inventories will rise. Note, we did not say your ideal inventory position. We will never reach the ideal. (NOTE: Since Lean views all inventory as waste, the ideal would be one piece flow into the process to produce the exact inventory needed at the right time in the right quantity to meet customer demand, no more, no less.) Right now we are trying to eliminate all of the fat (the waste) we have built into our inventory cycles, so we can see where the opportunities for kaizen lie.

Chapter 6

Using PFEP for Internal Planning

Once the Plan for Every Part (PFEP) is started, you can begin to use the new PFEP to help plan the internal logistics from your finish process to the finished goods supermarket loop. You will then plan your internal logistics process moving back upstream into the preceding loop from your raw materials supermarket to the line-side consuming process. You will plan each inventory loop until you have developed internal routes to support all the processes within your value stream. So, internal planning includes facility, cell, and supermarket layout, and internal route development. We also discuss the use of the PFEP to help us determine the best storage and delivery method based on usage and container quantity.

It is important to have a robust address system to assist the route operator in becoming more efficient. This will also help you understand what type of route is best for your facility: coupled or decoupled routes.

There are rules needed to effectively manage a supermarket and line-side flow racks. These rules help us design the type of storage areas you have in your supermarket. Not everything can go on a flow rack in the market, nor should it. For example: you would not want to say that everything has to be in a flow rack in the market, especially if you have high-use components. If you did, then someone would be loading them in the stocking side of the supermarket rack as quickly as they were being pulled from the picking side. The best option may be to deliver these in a full pallet to the line or stage the pallet in a pallet pick area in the supermarket.

Every part is associated with a cycle time for the route delivery and if you're handling a lot of small fast-use containers, then you're dedicating a lot of manpower to that one part, resulting in an overburden in routes, equipment, and manpower for that component. Route delivery is not a one-size-fits-all; you still have to think and listen to every part's story and make the best decision possible with the data you have. The more data you have, the better decision you can make. Finally, your supermarket design and location may change many times before you get to the final version. The goal is to have the purchased component supermarket as close to the receiving dock as possible and your finished goods market as close to the shipping dock as possible.

First and foremost in all of this is safety. Whatever we do, we have to be safe on the shop floor and dock areas while stocking, picking, and delivering components. In this chapter there are several guidelines to aid you in developing or expanding your safety initiative within your facility.

Internal Route Planning

Just as we have takt time and cycle times for manufacturing and distribution, we also need to plan how we will get components from the supermarket to the consumption points. The PFEP plays an important role in helping us see how this needs to work. Figure 6.1 shows the typical internal route loop of this portion of the value stream map.

Supermarket layout should be correlated with usage or consumption locations. This means that we will presort parts and store them according to where they are used so that we can quickly arrange to have them moved to the point of use. Our ability to receive, sort, store, and move to the consumption point will determine the location of the storage areas and the size of our supermarkets.

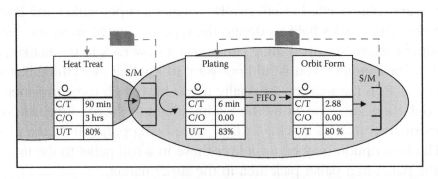

Figure 6.1   Internal loop.

Layout

When laying out your supermarket for timed delivery routes, your super-market should keep all the unique part numbers together based on the assigned delivery route and consumption point. This makes picking more cyclic. Common parts are kept separate to reduce the amount of total inven-tory that is required in the supermarket. For example, common screws, nuts, bolts, and cardboard may be used in different locations or cells in the facil-ity. Figure 6.2 shows a typical supermarket layout to accommodate unique and common parts areas, at the same time depicting two timed routes, one grey and one black. If you have multiple delivery routes in your facility, and if your delivery routes are "coupled," route paths will cross in the common parts locations in the supermarket.

Replenishment from the supermarket is performed by timed delivery routes from the supermarket location to the line usage point. These routes are either coupled or decoupled.

Coupled versus Decoupled Delivery Routes

A *coupled* delivery route is a route in which the driver delivers parts, picks up kanbans (signals) and empty containers, pulls replenishment parts from the supermarket location, and delivers back to the kanban usage point loca-tion on the next route cycle. Figure 6.3 illustrates the sequential steps the driver completes during a typical one-hour, or 3,600-second, delivery route.

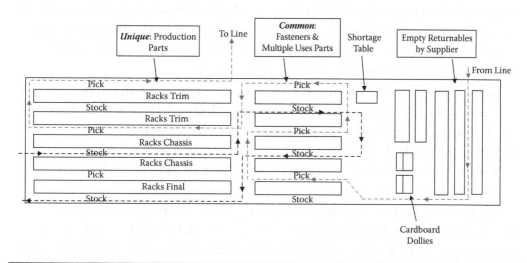

Figure 6.2 Dock layout flow.

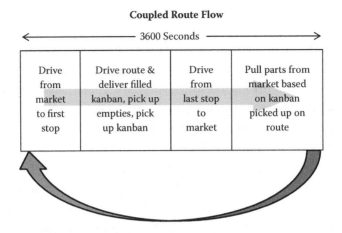

Figure 6.3   Coupled route flow.

A *decoupled* delivery route is a route in which the driver only delivers parts, picks up kanbans and empty containers, drives back to the supermarket, and unhooks from the delivery dollies. The driver then rehooks to staged full dollies that have been filled by a supermarket attendant from the kanbans turned in on the last cycle. Figure 6.4 illustrates the sequential steps in both the delivery route and the supermarket for the same one-hour, or 3,600-second, route.

Coupled routes require longer cycles between deliveries or more staff to do deliveries in the same amount of time. Decoupled routes require more kanbans (more inventories) in the system because the kanbans picked up on the current delivery route cycle will not go out until the next delivery cycle. Figure 6.5 shows you the difference in the kanban flow cycles of a coupled and decoupled route, from kanban pickup at the using location until that kanban is returned to the using location. You will have to do the PFEP calculations to decide which method works for you.

Be careful here; don't think decoupled is always better than coupled. The goal is to load the conveyance operator on a decoupled route at 85% maximum utilization on every trip. If that can be done in a coupled route, there is better control. If there is too much work content, decoupled is best.

Within your supermarket layout, you will need to know where to go to replenish the kanban. This is accomplished by an address system established to facilitate the stocking and pulling of those parts stored within the supermarket. Addresses are established and a visual marking system is put in place. This allows creation of standard work, timed delivery, and planned stops.

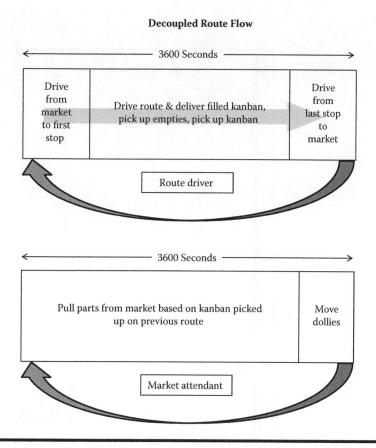

Figure 6.4 Decoupled route flow.

Address System

The dock layout is a living entity that evolves as the business changes. A logical numbering system is standardized and identifies unique locations on racks, conveyors, in off-line work areas, and in storage areas. This allows specific identified locations in the plant for delivering material; providing services; emergency response; and placing people, tools, and equipment. Figure 6.6 shows the typical address system flow.

Your standard of sort, set, shine, standardize, and sustain (5S) should require you to create a layout for your supermarket and identify where material will be received, stored, and shipped. The 5S allows us to create the basis for visual management and control to assist in identifying abnormal conditions, including abnormal (too much or not enough) inventory conditions.

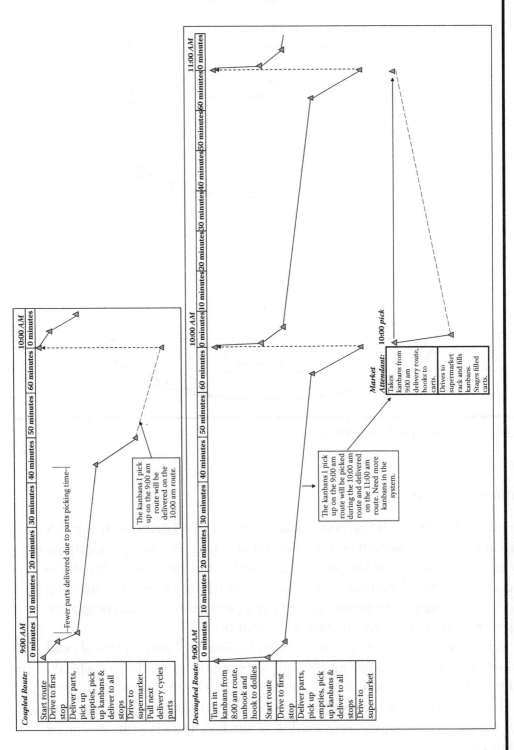

Figure 6.5   Decoupled route replenishment cycle.

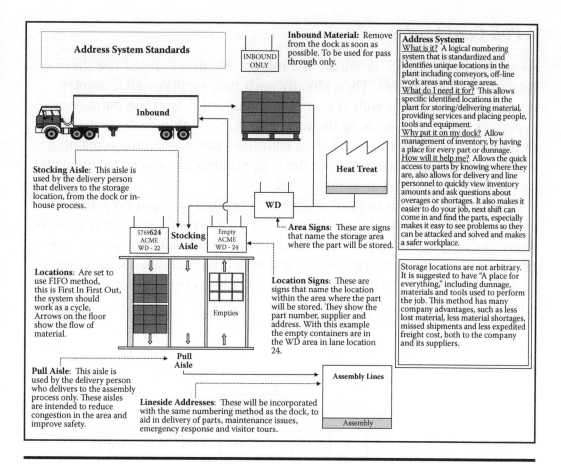

Figure 6.6   Address system standards.

Identify all material receiving, shipping, and dock locations. Create a visual control system by hanging signs that are easily updated, easy to install and replace when change is needed.

- Pull-Card Marketplace
- Call Marketplace
- Repack
- Overflow
- Industrial Materials
- Dunnage
- Scrap/Rework
- Pilot/Preproduction

A Pull-Card Market

As shown in Figure 6.7, designate the market title as a letter of the alphabet for each unique rack set. Then identify each rack section with a numeric designation, beginning with one, in ascending order. Start the numbering system as close as possible to the receiving dock. Within each storage rack or flow rack, identify the lower level with the letter A, then the remaining rows in alphabetical order. Within the row, number the space starting with one in ascending order as in Figure 6.8.

Standard row width is 12 to 18 inches. This unit is based on standard carton size, but you should be prepared in your layout with carton dimensions so that you can adjust as needed. In Figure 6.9 you can see we use grey and black electrical tape to show where a part storage location stops as a visual control if a single part number takes up two or more row spaces.

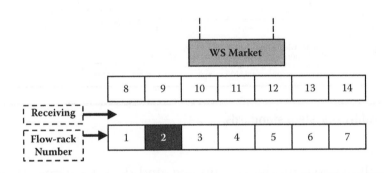

Figure 6.7 WS market.

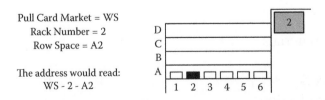

Figure 6.8 Flow rack.

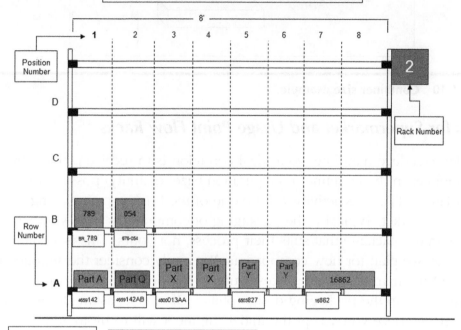

Part A Location = 2 -1A	Standards for Flow Rack Addressing:
Part Q Location = 2 -2A	
Part X Location = 2 -3A	
Part Y Location = 2 -5A	

Standards for Flow Rack Addressing:

1. Always start "Rack Number" in the left to right manner, each rack will have a set number. E.g., above is Rack 2.
2. Establish standard intervals for "Position Number, " simply put break the front length (presentation side) into equal size positions. In the example above, there are eight equal positions of 12 inches. This is done to establish a standard and allow for ease of adding and deleting storage positions without rearranging every item in the store.
3. Mark each location position with either red or green tape. This sets the standard of knowing exactly where a position ends and the next begins. Use the philosophy of a traffic signal, "red" stops the position and a new position starts on the right hand side of the tape. "Green" simply means that the box or area needed to store the determined quantity will not physically fit in the 12-inch space above, therefore requiring additional positions, so the "green" tape tells you the position "goes" until you reach the "red" tape stopping that part number.
4. Always mark the location with rack label; this label at minimum needs the following information: part number, part name, location (storage address), supplier. Set a standard for label placement; typically the label is place in the left-hand bottom corner of the position and next to the "red" tape. Keep this as uniform as possible, you should be able to walk to the rack and understand exactly what the standard is telling you and see the abnormal.
5. Only use one (1) label per part number placed in the standard position. If a part number requires more than one position on the rack for storage, place the label at the first position location only.
6. If more than one position is required, then you would move to the next position available in the position address to assign the next part. Ex. Part X location is 2-3A (its position takes up 3A and 4A); therefore, 4A would not show up in the addressing system as having a part number assigned to it. The system would show part numbers in 2-3A and 2-5A; this is done to allow the addition of a new part with ease when the space required by "Part X" no longer requires two positions.

Figure 6.9   Flow rack rules.

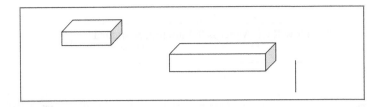

Figure 6.10   Container size example.

Rules for Supermarket and Usage Point Flow Racks

In order to reduce handling, your ideal condition is to receive product from your supplier in the container configuration (size, quantity) possible to meet Just-in-Time (JIT). This activity will be one of the final steps in setting up your supermarket. Typically, the packaging of components from your suppliers is in the package that suits their process, not yours. This is a failure in the way we plan for new components. We rarely consider the supplier as an arm of the manufacturing process but it is important that suppliers are involved early in the packaging design phase. Of course, asking for a packaging change once you accept the initial shipment will result in added cost to accommodate your request.

For the sake of simplicity we suggest that you pick three standard container footprints and three variations of those footprints; for example:

- ■ Footprint
  - – 12 in × 15 in
  - – 12 in × 22 in
  - – 22 in × 24 in
- ■ Variations (in height)
  - – 12 in × 15 in × 5 in
  - – 12 in × 15 in × 7 in
  - – 12 in × 15 in × 9 in

Choosing a simple approach such as this, with the only variation being the height simplifies storage rack and pallet configurations. See Figure 6.10.

Call Market

The Call Market consists of parts stored on the floor due to the size of the container or usage. When choosing your call market location, some factors you

Figure 6.11   WD call market sign.

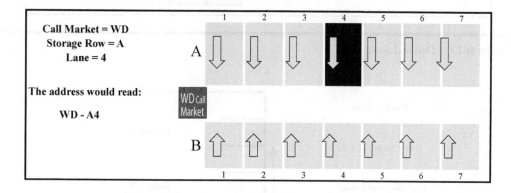

Figure 6.12   Call market layout.

should consider are distance from consumption point, distance from unload point, and whether the parts are unique to a cell (they should be grouped together) or common across multiple cells. Where possible you should locate supplier parts that come in on the same truck close to each other in order to speed the unloading of trucks and putting away of parts or components. Use the same principle for numbering the call market as the pull-card market. Use a sign over the area to denote the call market, as in Figure 6.11.

Identify each material's storage section, using alphabetical labeling starting with A closest to the receiving dock. Within each call market section, identify each lane, beginning with number one as shown in Figure 6.12.

Receiving/Shipping Address System

Identify on the plant layout all receiving/shipping dock locations. This is required to facilitate location identification and the quick loading and unloading of trucks. Docks should also have a unique identifier, as referenced in Figure 6.13 in your layout plan (i.e., blue, yellow, south, west). Assign each

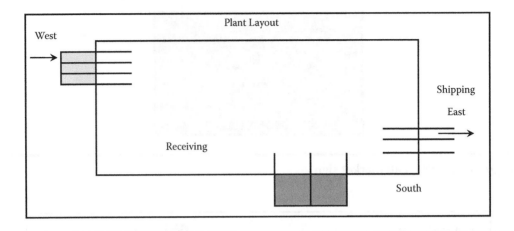

Figure 6.13   Dock doors.

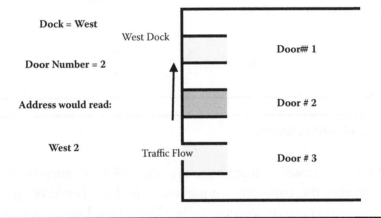

Figure 6.14   West dock doors.

door within each coded dock a number beginning with number one as shown in Figure 6.14. Signs should be mounted both internally and externally at each dock door, as shown in Figure 6.14, for quick reference by inbound trucks and dock personnel.

Other Areas

Identify the following addresses by name:

■ Repack
■ Overflow

Figure 6.15 Unsafe operation.

- Industrial Materials
- Pilot/Preproduction
- Dunnage
- Scrap/Rework

All plant systems should use the same address when referring to conveyors, work areas, and supermarket area. Maintenance and manufacturing control systems should also be able to support your address logic.

Within the facility, 5S rules should ensure that all signage is standardized for size, height, color, and markings. Signs should be visible from the aisles. Signs are normally hung at a height of 10–12 feet above the floor. Certain colors are easier to see, such as white or yellow lettering on a blue background.

Install "mall maps" at the main entrance to the facility and at major intersections. A mall map is a layout of the entire facility with all of the major areas identified. These are to be located in strategic communication areas and marked with "You Are Here" designators. The test of your "visual factory" signage is to enable a new operator to locate material or equipment without having to ask to speak to anyone.

Safety First—If It Is Not Safe, Do Not Do It!

All of us must ensure a safe and accident-free work environment to all employees. Our responsibility is to design a flow system to quickly recognize and eliminate safety concerns. This responsibility lies with everybody, but first with management. A workplace that is accident- and incident-free must be everyone's first priority; this has to be reflected in all areas of managing and operating a plant. All employees expect and deserve a safe and accident-free environment in which to perform their duties. It is imperative to take responsibility for safety and ergonomics.

Safety activities for materials areas are

- Only allow work-related people in the receiving, shipping, and supermarket areas.
- Quality control areas should be separated from vehicle traffic.
- Use receiving door dock locks or red/green light control.
- Carrier trucks must be safe for forklift unloading.
- Always use tacks and chocks or dock locks.
- Set height limits for material storage and along aisles; you should be able to see over the stored material.
- Store pull-card parts and heavy material at waist height in racks.
- Place stop lines at aisle exits.
- Set dolly heights to meet ergonomic requirements.
- Forklift trucks and tow motors require daily safety check sheets.
- Forklift and tow motor drivers need a licensing/recertification schedule.
- Schedule weekly safety audits with item/problem follow-up sheets.
- Train correct lifting and handling techniques for workers.
- Operation instructions and standardized worksheets must include safety precautions.
- Record unsafe conditions at once with a camera. Contain the problem and then look for the root cause.
- Never proceed without eliminating the actual danger of hurting someone.
- Implement an accident/incident/near miss tracking and monitoring system.
- No use of mobile phones while driving (calls or text messaging).
- No use of headphones of any kind (this does not include required hearing personal protective equipment).
- Loaded fork trucks should be driven with operators facing the rear, not trying to see over pallets.

- Must have an agreed-upon discipline process for those not following the rules (e.g., "unsafe forklift practice").
- Encourage everyone to contribute to prevention through a suggestion system or other means.

Recently we were doing training in a manufacturing plant. We broke the class into small teams and asked them to complete current state maps of specific processes. During the reports, one of the teams made the following observation "… and we observed a potential safety/ergonomic issue when an operator moved a mold and guided it with his hands, not the guide pole." We stopped the report to ask if any of them had spoken to the operator and supervisor.

The answer to both questions was no. We stopped the class and took them all to the production floor where the violation occurred. We met the plant manager and safety manager and asked the team to explain what they saw. We insisted that corrective action (not disciplinary action) be done while we were present. It took some time to get this done and this was "precious" classroom training time, but it is important to send the message that safety is our most important job, and it is everyone's job. It is in fact important enough to stop training and correct the problem before it becomes an incident or injury.

A manufacturing plant is a complex organization. For that reason we may require inventory supermarkets to buffer processes that run at different rates. Supermarkets may look and operate slightly differently, even within a plant. But the role of the supermarket remains the same; to provide a steady flow of parts and components to your manufacturing or assembly cells. It is important, however, that you have a simple and logical address system to help create and maintain a robust delivery scheme. The address system described above is very similar to the mail delivery system, and the delivery methods look a lot like a taxi or bus route system, with the bus being a lot more efficient than a one-passenger or a one-way taxi service.

As a business we owe it to our most prized resource, the people who make what we sell, to always put safety first. The address system is not just there to help with parts delivery; it has a huge safety role in supporting an emergency response should you ever have one. The more visual your plant, the easier it is for first responders to find the exact area to which they have to respond.

The next chapter builds on the delivery route by discussing the different types of part deliveries and what makes each one unique. It also covers how to set delivery cycle times and create the best route delivery frequency without overburdening the process.

Chapter 7

Delivering Parts to the Operators' Fingertips

The principle of Just-in-Time (JIT) is to have the right part in the hands of the customer, in the right quantity, at the right time. This same concept applies internally (operators are seen as customers of the inbound supply chain) as well as externally. The question we need to answer is how do we create the rules, methods, and processes for internal delivery? Internal route development needs to have its own cadence or takt time, so before we start any route development we need to decide not whether we use a timed delivery route but what the rules will be for developing our timed delivery routes.

The primary function of a timed delivery route is to replenish what has been consumed during the previous cycle. Material handling functions, moving material from one process to another, or moving finished goods to the shipping dock are secondary to the route and traditionally are performed by direct labor in the work cell. However, this creates inefficiencies and waste in your direct labor functions. Our studies in standardized work indicate that direct labor is 16 to 60% inefficient because the direct labor operators are performing noncyclic, non-value-added work. Moving the non-value-added activities out of the direct labor will improve your operating performance. There will be no more lost or wasted time while operators refill bins or search for equipment, parts, or tools.

Material handlers should be your best employees as they are carrying the DNA of your operation (your parts) and are setting the pace of manufacturing through the routes they run. This means you should take your brightest employees out of the manufacturing process to deliver parts. This is the most critical job in your facility. It requires the discipline of independent work with little direct supervision. It requires the operators to self-monitor their pace of production so that they keep the manufacturing cell running. It also requires them to think on their feet as changes or expedites will occur during the day. In short, it requires your brightest (best) employees to run these routes.

Equally important to designing timed delivery is developing rules for delivery triggers. With a timed delivery route these rules are as important as the run rules you have on the line. A rule that must be followed is that the timed delivery route has to leave at the scheduled time; it also has to return to the market when the route cycle is scheduled to end. There has to be an *andon* (help chain) point on the route, meaning the route has to be at a certain stop, at a certain time within the route; if it is not, then the route driver has to call or radio her team leader and ask for help. This may be asking for the team leader to start the next route at the scheduled time; then the route driver will catch up and finish the route cycle. Understanding these rules will help you implement a successful timed delivery route. Replenishment from the supermarket is done by timed delivery routes from the supermarket location to the usage or consumption point. There are two types of timed delivery:

1. *"Known Time—Unknown Quantity"*: You are delivering on a set timed route, but you do not know what has been consumed by the usage point until you pick up the kanban cards to tell you what to replenish.
2. *"Unknown Time—Known Quantity"*: With this method you are delivering and replenishing known quantities that have been consumed, but you have no set time pattern of the usage.

Within these two delivery types, there are three methods of delivery:

1. Small part (or) tote (pull method, multiple parts delivered in one cycle)
2. Call part (push method, one part delivered with one trip)
3. Sequence (requires double handling, avoid as much as possible)

Timed and scheduled delivery are key components of our internal JIT material process. To gain a deeper understanding of route development, let's look at the part types and how we can begin to design rules for setting up our delivery systems.

Small Part Delivery (*Known Time—Unknown Quantity*)

With small part delivery you are delivering smaller components to the usage point based on a calculation from the PFEP to match the timed delivery cycle. There are two principles that qualify a component to become a small part:

1. Percentage of daily build
2. Size, lightweight

Remember that with a timed delivery route, you are replenishing what has been consumed during a set route cycle. The loop between the usage point location and the supermarket is for only the amount of components required to build the cycle's worth of product. This amount is determined by the team that sets up the production cell and uses the PFEP to size the point-of-use flow rack to hold only the amount of each part number to support the process of the timed route interval. In this method the process is built around the packaging and weight standards. This allows us to keep components for multiple models or unit variation on the assembly line without requiring a material changeover. An assembly line or work cell will often be able to switch between models and options very quickly. If we have to wait to remove the previous model's component materials from the process and introduce the new model's component materials, we have added complexity and waste into the process. To gain flexibility and reduce waste, parts presentation must match the replenishment cycle and usage consumption. We have found that a good starting point for the replenishment cycle is a one-hour delivery cycle and two hours' usage stored at the assembly process.

At Toyota the goal was to have the container quantity at 10% of the day's run. As with everything at Toyota, there was a reason the 10% was selected. The 10% rule was in place due to the amount of fluctuation that could be absorbed at each assembly operation throughout the entire value stream from stamping until the vehicle reached final assembly without overburdening the operations along the way. Figure 7.1 shows that *heijunka*, or level and sequenced production, drove this decision for the percentage of variation in the packaging, supplier agreements, and line balance. The assembly processes were balanced to work content per model. For example, introducing three V6 engines back to back would overburden those jobs directly related to a V6 and allow the operation to go over cycle. The amount of fluctuation within your value stream will have to be determined by you and

Heijunka Pitch	1	2	3	4	5	6	7	8	1	2	3	4
Model	C	C	A	C	C	A	C	C	A	C	C	A
Engine Type	4 Cyl.	4 Cyl.	V6	4 Cyl.	V6	V6	4 Cyl.	4 Cyl.	V6	4 Cyl.	4 Cyl.	V6

C = Camry   A = Avalon   4 Cyl. = 4 Cylinder   V6 = 6 Cylinder

Figure 7.1 Heijunka pitch.

from that decision, the right percentage per container can be set. For Toyota 10% was a target; thinking was still required by assessing each part number's detail from the PFEP.

For parts delivered from the supermarket to the usage point, we want to set a weight standard per container. To set this standard, we begin with a target quantity per container of one hour's worth of components not to exceed a net weight of 29 pounds (components) and for this example, our empty container weighs 3 pounds for a gross weight of 32 pounds.

For example, if the daily PFEP takt time uses a build requirement of 450 units per day, then the recommended standard quantity per container target is 45 pieces using the 10% packaging rule. If the part weight is 1.2 pounds per piece, then the net weight of a 45-piece container would be 54 pounds, exceeding our weight allowance of 29 pounds. In this case the recommended standard pack would be 24 pieces, or 28.8 pounds net weight. In this example, the weight became the decision point because the 10% of shift/day run would make this container too heavy for the material handler.

Even with the best possible understanding of the PFEP data you will find yourself learning something new each time you perform the route planning process. A good example of how weight can affect an entire route occurred during the 1992 major model change at Toyota. I had used every tool and standard I had learned from my Japanese trainers to set what I had thought was the best possible route structure into place. The routes were balanced to the usage on the assembly line, operators were trained, and everything was implemented and running smoothly at startup. Then, about three months into the production model year, my Japanese trainer ask me why some tugger delivery routes were finishing early. Following the startup, I had spent time on the assembly floor and had observed a delivery route finishing early; however, I had attributed it to a short kanban cycle and did not stop to fully understand the process or what I had seen. Hence, I had not followed up on the progress of the routes. I soon discovered that a knack had been developed by the route drivers and a route time imbalance was beginning to show up. The thing about knack is that it cannot be taught but it can be learned, as we discuss in more detail later in this chapter.

What we had not discovered in the planning stages were the physical weight imbalance between the routes. The route structure adopted for this model change had been based on textbook thinking: calculating what is needed at each process, number of containers used per day, and the number of lifts performed at each station and then setting the length of the route, with the end result being the number of routes needed. Here lies the flaw, when I applied the standard out-of-the-book thinking: I had missed the total physical weight the routes were required to deliver over their entire day. Trim and final line delivery routes were delivering component containers that weighed far less than those routes delivering component parts to the chassis lines. Even though the component container weights per container were at or under the 29-lb standard, what I had not considered was the sheer volume and component mix across the different types of assembly lines. Simply put, the trim and final routes were handling on average 1.2 tons of components per day versus the chassis routes handling upwards of 8 tons per day. As was quickly pointed out to me by the chassis line route drivers: there's a lot of difference in handling sun visor covers than engine mounts.

The routes were redesigned to balance the workload and reduce the overburden by implementing routes that now serviced a trim, or final, line and a chassis line. There is no way to write a book to capture all of these variables: you will have to depend on the data you place in the PFEP, and you will have to go and see to observe the results of your plan to help guide you through each story every part is telling. This is the reason we only pick one finished goods part to focus on and follow through each of the processes when we pick a value stream to map. From there, we take that one finished goods part and begin to build the PFEP from its components within the Bill of Materials (BOM), learning everything possible about each of the components until we have every level in the part structure captured and understood.

Kanban Calculation Examples

Calculating internal kanban for the usage point, delivered by a timed delivery route, is as varied as the different types of processes that are scattered across industries. The main focus of calculating the internal kanban is understanding what is needed at each usage point location from the PFEP and at what rate those materials will be consumed. Internal kanban, as seen in Figure 7.2, will be used from the picking side of the supermarket flow rack to the usage point and only travel within this loop of the value stream. The goal should be to minimize as much decanting of components in the supermarket as possible, but at the same time we want to target either the hour's worth of consumption, 10% rule, or the weight standard. Receiving the correct container from the supplier is going to take some time to accomplish, due to most organizations' current packaging agreements.

$$\frac{\text{Customer Requirement}}{\text{Production Shifts for Line}} = \frac{\text{Line Demand per}}{\text{Shift}}$$

$$\frac{34}{2} = 17$$

Figure 7.2   Line demand calculation.

$$\frac{\text{Shift Available Time (min)}}{\text{Units per Shift}} = \frac{\text{Ideal Unit Cycle}}{\text{Time @100\% (min)}}$$

$$\frac{425}{17} = 25$$

Figure 7.3   Ideal cycle time calculation.

$$\frac{60 \text{ minutes an hour}}{\text{Ideal Unit Cycle Time (min)}} = \text{Units per Hour}$$

$$\frac{60}{25} = 2.4 \text{ (or) } 3 \text{ Units}$$

Figure 7.4   Units per hour calculation.

In this example for calculating internal kanban, the client builds air-conditioning units that attach to equipment control cabinets. The customer requirement is 34 units a day. The facility works two production shifts to produce these 34 units. This equates to 17 units per shift. Each shift length is 7.1 hours or 425 minutes of available time per shift (after all planned downtime has been removed from the 8 hours or 480 minutes) to produce, giving an ideal cycle time of 1,500 seconds (25 minutes) per unit as seen in Figure 7.3. At 1,500 seconds per unit, the demand on the kanban will be three components per hour as seen in Figure 7.4.

When planning, you would think that the build pattern would be as simple as dividing the 17 per shift units by the 7.1 hours available and getting the 2.4 or 3 units an hour. The truth is, the usage pattern for the shift is a mix of units per hour and not a 3, 3, 3 pattern. You will have to plan to the largest hour's consumption for the kanban usage, but there will be heavy and light delivery cycles resulting from the available time to build product.

To fully understand this, let's look at what makes up a shift's time. There is a 5-minute startup meeting in the 7 a.m. hour, a 10-minute break in the 9 a.m. hour, a 30-minute lunch break in the 11 a.m. hour, and a second 10-minute break in the 1 p.m. hour. Once you understand the pattern of usage, you can see the heavy and light deliveries on the off hours as shown in Figure 7.5. The same principle works if you're building only one unit every 1,500 seconds, one vehicle every 55 seconds, or one component every 0.06 seconds.

Understanding the Breakdown of the Product Mix

Now that we know what we are building, and at what rate, we need to understand the breakdown of the product mix. This will determine the pattern of the frequency at which the parts are delivered to the line. In the air-conditioner model from above, Figure 7.6 shows that the production floor is broken into cells based on the different models that are produced. Most lines focus on one or two model types. The air-conditioning line produces two different models.

Cell 5 builds two models of air conditioners. When we started to populate the PFEP for these two models the BOM showed that there was a combined variation of 683 different finished part numbers. The PFEP helps us to identify the common and unique parts between models and production lines, and it also helps us to identify the unique components for Cell 5's two models and allows us to determine which are standard and which are optional. On the PFEP under the column "Standard or Option" we code those parts that are used on all 683 part numbers as standard. An option is any finished part number that will use one version of a component on one finished part number and another version of that part number on another.

Example: Hypothetical finished part number "1652" has a solid door; finished part number "1670" is the same model family and has all the same internal components as "1652" but has vent holes in the door versus the solid door on "1652"; only one component makes it a different finished part number.

Cell 5 Usage Pattern versus Time

Hour and Times	1st Hour (7:00–8:00 AM)				2nd Hour (8:01–9:00 AM)				3rd Hour (9:01–10:00 AM)				4th Hour (10:01–11:00 AM)				5th Hour (11:01 AM–12:00 PM)				6th Hour (12:01–1:00 PM)				7th Hour (1:01–2:00 PM)				8th Hour (2:01–3:00 PM)			
Full Hour (mins)	60 Minutes				60 Minutes				60 Minutes				60 Minutes				60 Minutes				60 Minutes				60 Minutes				60 Minutes			
Available Mins.	55 Minutes				60 Minutes				50 Minutes				60 Minutes				30 Minutes				60 Minutes				50 Minutes				60 Minutes			
1/4 Hour	15	15	15	15	15	15	15	15	15	15	15	15	15	15	15	15	15	15	15	15	15	15	15	15	15	15	15	15	15	15	15	15
Start Up Mtg.																																
Unit 1																																
Unit 2																																
Unit 3																																
Unit 4																																
Unit 5 (Break)																																
Unit 6																																
Unit 7																																
Unit 8																																
Unit 9																																
Unit 10 (Lunch)																																
Unit 11																																
Unit 12																																
Unit 13																																
Unit 14 (Break)																																
Unit 15																																
Unit 16																																
Unit 17																																
Consumption Pattern (Units)	2				2				2				3				1				2				2				2			
Route Burden	Average				Average				Average				Heavy				Light				Average				Average				Average			

■ = Product Unit

■ = Planned Downtime (Breaks, Lunch, Start-up Meeting)

Figure 7.5 Consumption cycle.

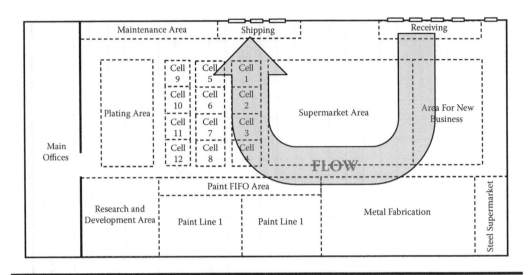

Figure 7.6 Overall plant flow.

Planning at the Cell Level

We are now ready to begin planning at the cell level. To facilitate and truly understand material flow in and around the cell, we have to have a cell footprint (Figure 7.7). At this point the part delivery method has to be established. As stated earlier, the Lean goal is to have all variations of a model presented to the operator to support any build variation that he or she may see.

From our PFEP of Cell 5, we found we had unique and common small component parts that we could deliver in totes, larger and heavier parts such as compressors that would need to be delivered as a call part, and mixed model or option parts that would make more sense being delivered as sequenced items due to the sheer variation.

We set the cell usage point quantities for components designated as small (totes less than 29 lb) at two hours inventory with a one-hour delivery cycle. This allows effective presentation of small parts, such as screws, nuts, and so forth, to support mixed model content variation at the usage point while still supporting an overall scheme of one hour delivery routes for parts with higher usage variation.

Lean Note: Presenting all required components to build any model or option within the work area or cell means that we cannot build waste into the process for walking to retrieve components. If the work area or cell has a frontage footprint of 12 feet, then all components must fit within that frontage footprint and not extend into the next process on either side.

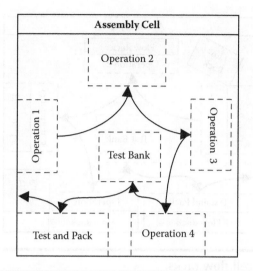

Figure 7.7 Assembly cell flow.

The PFEP identifies our standard common parts for Cell 5 and identifies container quantity and size for this cell. A one-hour delivery frequency will not produce a smooth, even flow of parts to the cells, because the takt time and flow of finished goods taken away from the cell will vary by hour due to breaks, lunches, and so on as represented in Figure 7.5. Now, we need to focus on the delivery of these parts to the cell.

Delivering the Parts to the Cell

Our air-conditioning client's process is a batch production process to support the order pattern of the customer: multiple models on one purchase order. The decision was made to be a "build-to-order" assembly line because the future state map showed that we could support shipment without the need for a finished goods market, reducing overall inventory by building in flexibility.

If the demand from the customer order requires model 1652 (no vent holes) production to run for three hours, then a short changeover takes place and model 1670 begins production. This means that the entire kanban quantity for 1652 unique components will end up back in the line-side flow rack at the end of the delivery cycle. Our cell flow racks will therefore need to be able to support not one hour of parts (our delivery cycle) but the entire three hours (four hours for decoupled route) of parts. Failure to plan for this would leave you with no place to store the parts. (You have this option or

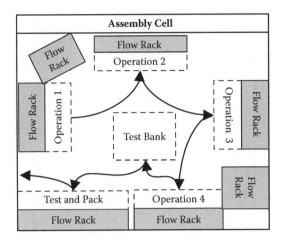

Figure 7.8   Assembly cell flow racks.

the waste of taking the parts back to the main supermarket; in some cases a rack to hold four hours of inventory becomes a size problem.) Figure 7.8 shows the flow racks that are required to support the point-of-use inventory to build the units for line five.

Figure 7.9 is our PFEP for the air-conditioner manufacturer location Cell 5. The highlighted line has a usage point kanban container quantity of 50 determined from our calculation of usage per. Each assembly unit gets four parts of part number 20-1014-10 (screw).

Our calculation shows that we should build an average of 2.4 units an hour to obtain our customer demand of 17 per shift. As an organization, we decided to deliver fasteners on average twice a shift, or once every 4 hours, to minimize waste in the supermarket during decanting of these screws. The calculation for a fastener is as shown in Figure 7.10

Think of it this way: a screw is normally very low cost, is received in bulk to reduce cost, but is used in small quantities and is subject to being dropped or cross-threaded at the usage point, making it hard to set a strict usage per container.

If you use the container as the kanban for fasteners, an easy method to keep from counting out each fastener is to draw a line in the container to give the material handler or supermarket attendant a quick reference point of the level at which to fill the container to meet the required consumption rate. For fasteners, it has been proven to be wasteful to have the route driver or supermarket attendant count out, for example, 50 pieces each time the replenishment is required. It is better to count that amount one time and mark the kanban container for quick reference. Remember, the route driver

Part Number	Part Description	Line Loc.	Store Loc	KB Qty
20-1014-10	Screw 10-24 X 1/2 HWH PF SED	C5 01	PT2	50
20-1014-15	Screw 10-24 X 1/2 HWH MOLD SED	C5 01	PT4	50
20-1014-214	Screw 10-24 X 3/8 TORQ HD ZP	C5 01	C14C4	50
20-1014-215	Screw 10-24 X 3/8 TORQ HD SS	C5 01	C10C5	50
20-1014-30	Screw 10-24 X 31/2 LOW TORQ HD ZP	C5 01	PT2	50
20-1014-35	NUT A KETS 10-24 ZP	C5 01	PT2	50
20-1015-04	NUT CAGE # 10-33	C5 01	PT2	50
20-1015-141	NUT A DIMPLE LOCKING 1/4-20 ZP	C5 01	C11C4	50
20-1015-20	WASHER FLAT 1/4 ZP	C5 01	C11C5	50
20-1015-64	WASHER FLAT 3/16	C5 01	C11C6	50
20-1015-76	WASHER LOCKING INT #10 ZP	C5 01	PT2	50
20-1015-92	WASHER LOCKING INT #10 SS	C5 01	PT2	50
20-1015-97	WASHER LOCKING INT #20 SS	C5 01	PT2	50
20-1047-02	CPLG BUSHING FLUSH 3/8X1/4 OD	C5 01	C11B3	3
20-1048-01	CPLG ROLLED STOP CXC 1/4 ID	C5 01	C11B4	6
20-1050-01	WIRE TIE 5-7/8i 18 Ib	C5 01	C11C2	9
20-1053-03	WIRE TIE, HIGH TEMP, 19 X 7.4	C5 01	C11C3	12
20-1053-14	WIRE TIE, HIGH TEMP, 22 X 9.5	C5 01	W16D3	3
48-2023-07	ELEC BOX STRN 1120 FLAT	C5 01	C14C2	6
62-6048-46	GROMMET SNAP PLUG DD. 875 OD	C5 01	C14C3	3

Figure 7.9   Assembly operation 1 PFEP.

9.6 Units in 4 Hours × 4 Screws Each Unit = Total Screws Used In 4 Hours
9.6 × 4 = 38.4 (or) Round Up to 39

Figure 7.10   Total hourly usage.

has to be the most efficient person in the plant because the operation is pure waste, and you do not want the route driver spending minutes having to manually count or weigh out on a scale 50 parts that cost 1¢ each.

We consider the route driver operation to be pure waste because the picking and delivery process (as well as the put-away) would not be required if the parts came off the truck and were delivered directly to the usage point synchronized to the rate of production. Other than this scenario, all of the other material handling functions we perform do not meet the strict definition of value added. But just because material handling is waste does not mean that we should not try to make route delivery efficient.

In our Lean simulation, we demonstrate the importance of route development and container planning from the supplier in round two of the simulation. This round goes much better than round one. But now, in round two, material flow becomes the bottleneck process, often holding up production as mixed parts arrive at the line or wrong parts arrive, or operators are usually waiting on delivery of parts. This is where the participants begin to understand the importance of material flow in the manufacturing process.

Calculating Delivery Frequency

Let's now examine the remaining parts in our air-conditioning example using our PFEP. Figure 7.11 shows the linking of the PFEP for the cell workstations one and three and the component part numbers required at each workstation to support the functionality build order of the air-conditioner.

With our usage defined in our PFEP, we can now begin to calculate our delivery frequency by part (kanban delivery), and the number of routes we need to support our production plan. Figure 7.12 shows a typical route flow from the supermarket, through the cells on the route and back to the supermarket. In this example we have decided that one route driver can cover Cells 1 through 6. Before we repeat the calculation we did in Figure 7.10 for the remaining parts for each delivery point we have a few more planning factors to consider.

The cycle for a timed delivery route is the established delivery cycle time; in our example we have chosen one hour and we want to balance that route to meet that cycle time. We generally want to utilize the route driver at the same efficiency rate (in our example, 95%) as the line operator. We are concerned with the number of lifts (or kanbans) delivered in a cycle. It will be inevitable that the route driver will get the knack of handling more than one lift at a time at a delivery point. When this happens, the route driver will become inefficient and waste will occur. Think back to our earlier story of knack and weight overburden and you will see how this is all tying together. When Toyota designs routes, they set the utilization at 100% to balance the

Part Number	Part Description	Line Loc.	Store Loc	KB Qty
20-1014-10	Screw 10-24 × 1/2 HWH PF SED	C5 01	PT2	50
20-1014-15	Screw 10-24 × 1/2 HWH MOLD SED	C5 01	PT4	50
20-1014-214	Screw 10-24 × 3/8 TORQ HD ZP	C5 01	C14C4	50
20-1014-215	Screw 10-24 × 3/8 TORQ HD SS	C5 01	C10C5	50
20-1014-30	Screw 10-24 × 3 1/2 LOW TORQ HD ZP	C5 01	PT2	50
20-1014-35	NUT A KETS 10-24 ZP	C5 01	PT2	50
20-1015-04	NUT CAGE # 10-33	C5 01	PT2	50
20-1015-141	NUT A DIMPLE LOCKING 1/4-20 ZP	C5 01	C11C4	50
20-1015-20	WASHER FLAT 1/4 ZP	C5 01	C11C5	50
20-1015-64	WASHER FLAT 3/16	C5 01	C11C6	50
20-1015-76	WASHER LOCKING INT #10 ZP	C5 01	PT2	50
20-1015-92	WASHER LOCKING INT #10 SS	C5 01	PT2	50
20-1015-97	WASHER LOCKING INT #20 SS	C5 01	PT2	50
20-1047-02	CPLG BUSHING FLUSH 3/8×1/4 OD	C5 01	C11B3	3
20-1048-01	CPLG ROLLED STOP C × C 1/4 ID	C5 01	C11B4	6
20-1050-01	WIRE TIE 5-7/8i 18 lb	C5 01	C11C2	9
20-1053-03	WIRE TIE, HIGH TEMP, 19 × 7.4	C5 01	C11C3	12
20-1053-14	WIRE TIE, HIGH TEMP, 22 × 9.5	C5 01	W16D3	3
48-2023-07	ELEC BOX STRN 1120 FLAT	C5 01	C14C2	6
62-6048-46	GROMMET SNAP PLUG DD. 875 OD	C5 01	C14C3	3

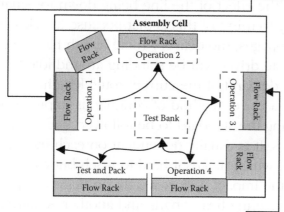

Part Number	Part Description	Line Loc.	Store Loc	KB Qty
07-3028-203	HARN W/L, AKC, M33, TS/ETS	C5 03	G10E1	6
38-2301-00	BLOWER MOUNT 412 W/TOP FLANGE	C5 03	G10E1	6
22-2302-00	BLOWER MOUNT TPR55 (I/O)	C5 03	G11B1	3
88-9903-01	SEALANT, NON-SILICONE	C5 03	G11C3	12

Figure 7.11   PFEP matched to operation.

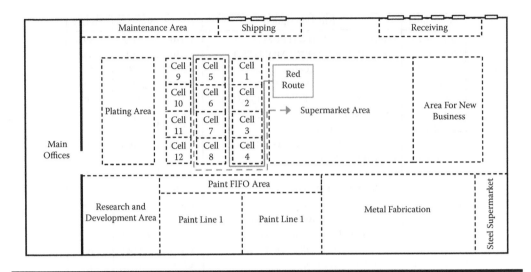

Figure 7.12   Route flow.

knack effect and also the effect of the line being down for small increments of time. The easy way to think of this is that because knack will occur and the line will have stoppages, the delivery route length has to accommodate the variable or the route driver will be standing around idle because he has met the consumption demand of the current route length.

Lean Note: Knack is a clever method of continuous improvement learning by an operator allowing the standard to be followed but minimizing waste and easing the process. It is usually dependent on each individual's thought process and dexterity, which makes it hard to teach but easy to learn.

We next consider the impact of part weight on delivery cycles. If one material handler is delivering heavy parts and another is delivering smaller, lighter parts, our experience tells us that the one with the light boxes will move faster than the one with heavy boxes. We want to balance light and heavy as much as possible to keep the operators working at the same pace.

Calculating the Number of Kanbans Delivered

We next cover the calculation to determine how many kanbans may be delivered by a delivery route driver. Even though we are setting a delivery cycle of one hour, there is a total number of kanbans that can be efficiently delivered throughout the entire shift by each route driver. This number will establish the path, length of the route, and total routes needed to support production.

At this point there will be two scenarios for determining the number of kanbans that can be delivered:

1. You are implementing timed delivery routes for the first time in your organization.
2. You currently have delivery routes but need to understand how to increase their efficiency.

If you do not have delivery routes, the method to determine how many kanbans can be delivered is going to be trial and error at first for the simple reason that there is no history, only planning.

We typically start to gather data from the PFEP from one pilot cell to deliver kanbans to, from our first attempt at a supermarket supporting the one-hour delivery route. At first, when you start this timed delivery, the driver will be inefficient.

The rate at which kanbans move through the cell is easily calculated: begin with the total number of part numbers and the usage per hour from that container, giving you the total number of kanbans to be delivered on a typical route cycle. At this point we can begin to calculate the distance to and from the new supermarket location, realizing the farther the driving distance is from the supermarket to the pilot cell, the less time the delivery driver has to actually handle and deliver replenishment kanbans, which could result in additional delivery routes. A good rule of thumb is that the driving distance to and from the supermarket should not exceed 33% of the total route time or, for a one-hour delivery route, 20 minutes total round trip.

The layout out of the supermarket is just as important as the driving distance. You do not want the route delivery driver spending time in the market looking around and driving back and forth to find part numbers. This is where we use the PFEP columns "unique" and "common." The supermarket should have been set up to support all the unique items for that cell to mirror the flow of the work cell for ease of picking and building the delivery route. At this point the common parts should have been identified and placed in a manner to allow for ease of picking; however, we do not recommended that these parts be placed and managed by numerical order due to changes, addition, and deletion of component part numbers. Remember, a supermarket is managed by location, not part number; we are more concerned about the number of locations and their support factor to the delivery route than we are of what is actually in the location.

If we have understood the part number to be replenished, drive time, and the supermarket layout, we can now begin to track the true number of kanbans delivered on each route and the length of time the route is in operation. We graph this information to begin the process of building a history. Figure 7.13 shows a tracking graph format to begin this process. As new cells are restructured and added to the timed delivery route, you begin to see the original route becoming overburdened. At this point you have to start using the history data to rebalance the route, possibly adding an additional route, moving the supermarket, realigning the supermarket, and so on. Once the history is built, it becomes very easy to use this information to see the waste present in the delivery route and supermarket. This entire materials delivery process is a living entity and cannot be set up once and forgotten.

Making Your Routes More Efficient

Now let's look at how we take the original delivery route or current routes, and make them more efficient. The data from the original or current route(s) will help determine the "time per kanban." Use a route audit sheet as in Figure 7.14 as a guide to begin gathering information to start the calculations. An important note here, remember the route driver needs to take breaks and lunch at the same time as the line they are supporting.

1. What is the available time the route driver has in a shift to deliver kanbans?
   a. For this example, our shift is 8 hours in duration.
   b. To find the true available time in our shift, we want to start with our 8-hour or 480-minute shift and deduct any planned downtime such as breaks, lunch and planned startup meetings.
   480 shift minutes
   − 20 minutes total covering two breaks
   − 30 minutes total for one lunch
   − 5 minutes total for shift start up meeting
   = 425 minutes available to deliver kanbans
2. Convert the 425 available shift minutes to seconds.
   a. 425 minutes × 60 seconds/minute = 25,500 seconds available per shift
3. Use the actual information from the route audit sheet to determine the delivery time and how many actual kanbans were delivered on each hourly delivery cycle. You will notice in Figure 7.15 there are hours where fewer kanbans are delivered than other hours. This decrease is due to the hours were breaks and lunch are present.

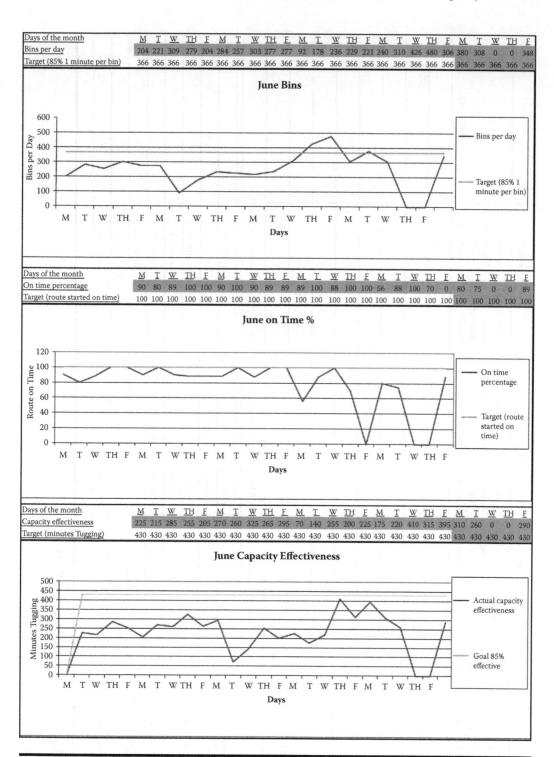

Days of the month	M	T	W	TH	F	M	T	W	TH	F	M	T	W	TH	F	M	T	W	TH	F	M	T	W	TH	F
Bins per day	204	221	309	279	204	284	257	303	277	277	92	178	236	229	221	240	310	426	480	306	380	308	0	0	348
Target (85% 1 minute per bin)	366	366	366	366	366	366	366	366	366	366	366	366	366	366	366	366	366	366	366	366	366	366	366	366	366

June Bins

Days of the month	M	T	W	TH	F	M	T	W	TH	F	M	T	W	TH	F	M	T	W	TH	F	M	T	W	TH	F
On time percentage	90	80	89	100	100	90	100	90	89	89	89	100	88	100	100	56	88	100	70	0	80	75	0	0	89
Target (route started on time)	100	100	100	100	100	100	100	100	100	100	100	100	100	100	100	100	100	100	100	100	100	100	100	100	100

June on Time %

Days of the month	M	T	W	TH	F	M	T	W	TH	F	M	T	W	TH	F	M	T	W	TH	F	M	T	W	TH	F
Capacity effectiveness	225	215	285	255	205	270	260	325	265	295	70	140	255	200	225	175	220	410	315	395	310	260	0	0	290
Target (minutes Tugging)	430	430	430	430	430	430	430	430	430	430	430	430	430	430	430	430	430	430	430	430	430	430	430	430	430

June Capacity Effectiveness

Figure 7.13   Delivery route performance.

Route Number:		Shift:		Date:			Operator:	
Planned Route Start Time	Actual Route Start Time	Planned Route Stop Time	Actual Route Stop Time	Variance (+ / −) (minutes)	Actual Route Time (minutes)	Number of Kanban Picked up for Replenishment	Number of Kanban Delivered on Route	Number of Shortage Kanban (unfulfilled)	Comments
7:05		7:59							
8:00		8:59							
9:10		10:09							
10:10		11:00							
11:30		12:29							
12:30		13:39							
13:40		14:39							
*2:40		15:00							

*Last pick of the day; replenishment of previous kanban route should be pull from market and left of carts for next shift delivery cycle start.

Shortage Part Number:	Delivery Location:	Comment:
Shortage Part Number:	Delivery Location:	Comment:
Shortage Part Number:	Delivery Location:	Comment:
Shortage Part Number:	Delivery Location:	Comment:
Shortage Part Number:	Delivery Location:	Comment:
Shortage Part Number:	Delivery Location:	Comment:
Shortage Part Number:	Delivery Location:	Comment:
Shortage Part Number:	Delivery Location:	Comment:

Figure 7.14   Delivery route tracking sheet.

Time Delivery Route (Time vs. Kanban Delivered)											
Delivery Cycle (at 1 Hour Delivery Frequency)		*1*	*2*	*3*	*4*	*5*	*6*	*7*	*8*	*Overtime*		Total
										9	*10*	
Planned Shift Delivery	(mins)	*55'*	*60'*	*50'*	*60'*	*30'*	*60'*	*50'*	*60'*	*0*	*0*	425'
	(secs)	*3300"*	*3600"*	*3000"*	*3600"*	*1800"*	*3600"*	*3000"*	*3600"*	*0*	*0*	25,500"
Actual Delivery Time	(mins)	54'	63'	46'	49'	30'	58'	44'	53'	0	0	397'
	(secs)	3240"	3780"	2760"	2940"	1800"	3480"	2640"	3180"	0	0	23,820"
Actual Number of Kanbans Delivered		72	88	91	61	89	47	81	66	0	0	595
Seconds Spent per Kanban		45"	43"	31"	49"	21"	75"	33"	49"	0	0	Average 44"

Figure 7.15 Total time and kanbans delivered.

4. Determine the seconds spent per kanban, within each delivery cycle, by dividing the hourly cycle seconds by the total kanbans delivered during that cycle.
5. Find the average seconds per kanban delivery, by taking the average of "Seconds Spent per Kanban." This becomes your basis for planning routes, purchasing equipment and planning manpower.
6. To find your utilization of this delivery route, take the actual delivery time and divide it by the planned shift delivery time.
   a. 23,820" / 25,500" = 93.4% Utilization

This section is about making your routes more efficient and to carry this thought further do not stop and accept that an average of 44 seconds per kanban, or 93.4% utilization is good enough. Take this data and question why you are spending 44 seconds per kanban; is your supermarket too far away from your assembly operations, is the supermarket not laid out in the correct manner, is the supplier container causing you to decant/repack adding time to the delivery driver, etc.? All these items, and issues add to the time required for the delivery driver to get a usable product to the user's fingertips. To give you a benchmark, Toyota in Georgetown, Kentucky used 29 seconds per kanban as a target for planning their routes. However, they already had the advantage of JIT, correct containerization, structure and sustainability on their side.

This exercise of understanding the number of kanbans that can be handled per route driver is crucial to helping you set manpower, material handling equipment, racking, container types, and other material handling needs, especially in an environment such as a vehicle assembly operation, such as Toyota, that deals with minor and major model changes annually. You now have a method to rebalance your material flow when demand changes or fluctuates due to new business increases as well as downturns. It's really as easy as monitoring the kanbans delivered and performing the calculation again.

Call Part Delivery (*Known Quantity—Unknown Time*)

Large components or containers require a different trigger for movement and method of delivery. We call these "call parts" because they are not on a scheduled delivery time; rather, delivery is scheduled when the operator "calls" for more. The call part is a known quantity and unknown time delivery. You will need some method to call for replenishment before the operator runs out or to remove finished goods from the cell. For the safety of your employees, you should attempt to keep all forklift traffic off the manufacturing floor, but sometimes that is not possible. If you have to move pallets of components, use walk-behind forklifts where possible.

If you do require call part delivery within your production area, make sure that you utilize the forklift in an efficient manner by carrying either a full or empty container on your forks at all time. At Toyota you are only to move with empty forks at the beginning of the shift and that is only from the forklift parking area to the call monitor where you picked your call ticket (kanban) transmitted from the line and then to the supermarket location to pick up your first delivery. When you made your delivery to the line-side location and swapped out the full for empty, you drove back to the call monitor with the empty pallet, picked your next call ticket, drove to the empty staging area, dropped the previous call tickets, emptied the container, and then drove to the supermarket location for the next delivery. This method allowed for extremely efficient operation of a process that in other companies may at best be 50%.

Call part delivery drivers have cycle times just as the timed delivery route. Figure 7.16 shows you how to calculate the cycle time and how it can justify the total manpower and equipment needed to support the assembly operation. As with small parts, distance traveled is a huge factor in calculating and establishing a cycle time for this method of delivery.

At Toyota, we used a dedicated computer-controlled call system. This system was linked to a central call monitor that input each part number and location manually and required engineering and maintenance support, at times becoming expensive to maintain. Your call system can be as simple as a light system, radio, or pager carried by the forklift operator. Just make

Call Part Manpower Calculation

1. Determine feet per second traveled at 7 mph.

$$\frac{5280\,\text{ft}\ /\ \text{mile} \times 7}{\text{Seconds}\ /\ \text{hr}} = \text{Ft travel in one second}$$

$$\frac{36,960}{3600} = 10.28\,\text{Ft}$$

2. Determine time to drive from column to column.

$$\frac{\text{Distance Between Columns}}{\text{Ft traveled in one Second}} = \text{Drive time between columns}$$

$$\frac{40}{10.3} = 3.88\,\text{Seconds}$$

3. Determine time to drive round trip from supermarket.

Total Roundtrip Column × Drive Time Between Column = Total Drive Time (seconds)

$$17 \times 3.88 = 66$$

4. Determine number of lifts per delivery. (*30 seconds per lift, allows for unwrap, exchanging Dunnage, etc.*)

   a. One lift in supermarket to set down previous empty into supplier empty location.

   b. One lift to pickup part to be delivered to call usage point

   c. One lift to set down full at call usage point.

   d. One lift to pick up empty at call usage point.

   e. One lift to pickup full and place into location at call usage point.

   f. One lift to pickup empty at call usage point.

   Total Lifts × per Lift Second = Total Lift Seconds per Delivery

   $$6 \times 30 = 180$$

5. Time allowed to get call ticket, if you are using a electronic calls (or) if using a radio time to get the call and log the delivery. Typically we use 30 seconds.

6. Total of all elements of a call delivery.

   Get call and log or get ticket = 30 Seconds

   Total drive time for call   = 66 Seconds

   Total lift time for call   = 180 Seconds

   Total   276 Seconds per Delivery

Figure 7.16 Call part manpower calculation.

7. Determine Number Deliveries an Operator Can Make in a Shift.

$$\frac{\text{Shift Available Time}}{\text{Seconds per Delivery}} = \text{Delivery per Shift}$$

$$\frac{27,700}{276} = 100.3$$

8. Number of Call Operators Required per Shift.

$$\frac{\text{Total Required Deliveries}}{\text{Deliveries per Shift}} = \text{Operators Required}$$

$$\frac{16}{100} = .61$$

Figure 7.16   (Continued)

sure you have a planned system to ensure that the line does not run out of components.

Sequence Part Delivery (*Known Time—Unknown Quantity*)

Sequence part delivery should be your last resort to deal with material due to the double handling. A part becomes considered for sequencing when the size, shape, variation, and color are so great that placing every variation at line-side would result in the operator walking a greater distance to obtain parts. With this also comes the floor space required at line-side to store the variation.

Some clients we have worked with have had to implement sequencing to create flexibility of model or color mix. In an example, we worked with a client that had a total of three model variations on one line, with a range of 11 colors across those models. The optimal batch quantity per color through the paint system was 10 individual parts requiring the same color to justify a paint color changeover. In the past, they had set their production and paint schedule to run three colors a day. Simple math will tell you that there is no flexibility or customer demand in this model. To get around to all 11 colors at only three colors per day was taking the system 3.6 days and each color change required a material change at assembly line-side. Sequencing offered them the flexibility to run every color every day, gaining them quicker lead time to the customer, reducing finished goods inventory, and allowing them to level out their schedule and not pull ahead to try to get the required 10 parts of the same part number to run their paint system effectively.

There are more variables than we could possibly begin to list in helping you determine the type of route planning you need. If you structure your PFEP correctly from the beginning to help answer the points we have made in this chapter, then the task becomes much more manageable. Setting routes and delivery methods is not a once and you're done endeavor but a daily activity that you will need to measure and countermeasure with each route run, call part delivered, and sequenced item staged.

You should now begin to see why an efficient system requires more than simply pulling an operator out of the process, giving him a tugger and cart, and telling him to start picking up cards and then deliver parts. The steps in route planning are as complex as anything you will do, but there are some key considerations. Will you run a fixed cycle and vary quantity based on consumption patterns, or will you vary the cycle time and deliver a known quantity? How will you design the delivery routes to make your routes and material handlers efficient? Finally, how does the structure of your BOM and your PFEP influence your decisions in route design? We wish there were one simple "cookie cutter" that would apply to every situation, but the complexity of your product mix and your manufacturing operations will dictate the right methods for your value stream. Every part has a story, and you have to listen to that story before you can develop the logistics part of your PFEP.

Chapter 8

Planning: Supporting Processes

Up to this point we have talked about what it takes for you to get your processes streamlined, possibly in cells, and minimum inventory online, pulled by kanban from a supermarket and delivered on a timed delivery route. This works well if you are in an assembly line environment; however, we have not talked about supporting processes such as planning, purchasing, stamping, metal fabrication, welding, painting, and so on. The next few steps are crucial in developing a total Lean enterprise. We say this because most Lean journeys stop here for the simple reason that up to this point it has been rather easy to implement, but now is when you will start to see the return on your investment start to take shape.

There are usually four points in time where you will truly start to see the gains in transforming your operation:

1. When you first start line balance and you remove all unnecessary waste and material from the producing processes
2. When you start to plan based on consumption and capability of the producing process
3. When you begin a good total productive maintenance (TPM) and changeover process on your machinery
4. Finally, when you truly begin to utilize supplier pull

The first three are completely under your control at the speed to which you furnish resources and dedication to the process. Unfortunately, we typically find that an organization is satisfied with just getting standard work, balancing a line to get manpower out, implementing somewhat of a kanban system, and forcing all the work in process (WIP) back into a so-called supermarket, all because the Lean journey is treated as a project. This is major surgery for your organization and should be treated as such; the process has to be allowed to move to the next level and follow-up is a must.

The fourth step is the hardest because it deals with your suppliers getting on board with you and truly becoming an extension of your value stream, trusting the new process, and seeing a true benefit to changing their organization to match the rate of pull you will demonstrate as your processes level out. A good rule of thumb here is that it will take you three to five years to get to this point. This is because you will need to get your internal system set and demonstrate a stable pull to the supplier with minimal fluctuation in demand. (Remember, your customer will not level your schedule. You will have to do this, and the processes have to be in place to allow for such leveling.) The main reason is in the packaging; most suppliers ship to you in quantities that are best for their processes, not yours.

Modeling Our Scheduling Process

There are two types of business model processes that will help you get to the point of scheduling the organization to be able to build the next step in the process:

■ True assembly based
■ Batch-supporting processes

When we talk about true assembly-based processes, we are talking about those that have a fixed line and the value-adders are the people who actually assemble components into a finished product such as cars, motorcycles, or toasters, among other things.

A batch-supporting process is a process where the machinery is large, changeovers are long, hits are fast, the machine is the value-adder, and manpower is indirect or non-value-added (i.e., screw machines, injection molding, forging, stamping, casting, extrusion, cold heading, robotic welding, etc.). The latter is the majority of the processes we find in the manufacturing environment and the least touched by Lean to date. It is also the most effective place to see gains in your investment down the Lean path.

Just-in-Time Scheduling

As we work through the process of Lean planning, we have to recognize that we will move you away from using your material requirement planning/ enterprise resource planning (MRP/ERP) system to plan day to day. We still have to have a robust MRP/ERP system but for forecasting demand to the supply base to ensure lead time and as a strategic in-house planning tool.

Let's recap what we have read and start to show why Just-in-Time (JIT) is so important to helping us schedule our value stream.

Just-in-time refers to producing and conveying what is needed, when it is needed, in the exact amount needed. Its attempt is to produce with either no or an absolute minimum of in-process inventory, resulting in a short-ened lead time and substantial savings in carrying cost, building quality, and worker motivation.

If you look at a traditional production system, it has several thousand part numbers required to be at the correct process, or hundreds of processes, at the exact time for assembly. Some of these parts require scheduling months in advance of the day of build. So a detailed schedule has to be drawn up and distributed to suppliers, and production can move forward accordingly.

However, it is almost impossible that things will go as planned. Conditions not known at the time the plan was created are bound to come up, and often drastically so. Even if changes do not occur, it is overwhelming to coordinate the thousands of parts and hundreds of processes with a large model–option combination and account for the problems that will inevita-bly develop during the manufacturing process itself at the same time. This results in inventory buildup and lead times becoming increasingly longer.

Just-in-time is designed to avoid all the above pitfalls by incorporating three operating principles:

1. The pull system (as accomplished by kanban)
2. Continuous flow processing
3. Takt time (synchronized processing speed)

Continuous flow processing and takt time shorten the production lead time through the guarantee of synchronized processing throughout the manufacturing process. The pull system as accomplished through kanban is the controlling mechanism that prevents overproduction and assures prompt and accurate dissemination of needed information.

These three principles have a necessary precondition in order to operate effectively: *heijunka*, or leveled and sequenced production. Heijunka means averaging both the volume and sequence of different item or model types on a mixed-model production line. Heijunka is the aim of the Toyota Production System's (TPS) planning and control. It is important that you understand the role of pull, continuous flow, and takt time to fully understand heijunka.

If you have been around different worksites, you have probably seen production levels jump to peak capacity and at other times drop to low volume, and on some days the number of units produced is more or less stable. Changes in production volume tend to cause waste, and the more frequent the variation, the more waste is created. On a line where the volume keeps changing during the month, the production capacity must match the peak loads, meaning that the worksite must always have enough capacity (machinery, materials, and manpower) on hand to react to the largest work order. On the other hand, if customer demand is lower than the peak level, the machines will not always be in use, the workers and machines will have idle time, and parts are more than likely to pile up.

TPS utilizes leveled production to avoid this wasted capacity or unevenness. The first step to TPS is setting an average for the production volume. If this average figure can be maintained for each worksite, department, and so on, then you can run the right number of processes with the right number of people every day. Failing to find this balance means that you have to have capacity to meet highly variable demand. If peak demand does not appear, we will be tempted to produce stock that is not required by customer demand, just to keep machines running and people busy.

True Assembly-Based Production

You have probably heard people say that all you have to do is level your production. With an assembly line you can do just that. The understanding of the model mix and the complexity of the model is the key here. Anyone you talk to from Toyota in Georgetown, Kentucky, will be able to tell you the mix in plant one is "1 in 3": one Avalon then two Camry. But there is another layer for the assembly department that stems all the way back to writing the standard work: work balance and planning manpower. This item is the engine for the assembly line.

The rule was that within the mix of 1 in 3 you could only process two V6 engines back to back without causing the line to suffer by overburdening it. The planning here is very critical inasmuch as a Camry can contain either a four-cylinder or a V6 engine; all Avalons were V6. In order to prevent the creation of batch processes, the flow was limited to 1 in 3. What happens when you cannot control your mix as tightly as Toyota does? It requires you to create batch processes to support your operation.

Batch-Supporting Process

Batch-supporting processes are the form of manufacturing with which you will most often come in contact. Being able to understand their demand and plan for it gets confusing because a batch process usually has several customers downstream and all want something at the exact same time. We need to understand that many manufacturing processes—stamping, injection molding, and painting—are fundamentally batch processes, but that does not mean that finishing and assembly are also batch processes.

We want to understand this mix, with its variation, and break it down into manageable elements. The biggest hindrance to allowing this to happen is the way MRP tries to schedule the machine. With MRP, one time you will receive a quantity of 10,527 to run for a part number, another time for the same part number the quantity will be 3,591, and the next time 26,104. There is no way in that scenario to plan stability in the process, nor can you accurately measure setup time, true scrap, and so on. The idea is to study the machine and its components and understand what sets the best run pattern to allow you to use the PFEP, setup, and production wheels, A, B, C, D coding (by volume), and supermarkets. If we have all of these elements we can establish our Every Part Every Interval (EPEI). Once we understand these, we can in time determine the EPEI and what we can work on to decrease buffer inventory. The three items we focus on from the Dynamic Capacity Planner (DCP) are uptime factor, setup time, and container quantity.

We use a DCP tool to plan batch-processing machines by first understanding:

- Customer demand
- Material requirements
- Tooling requirements
- Container requirements
- Setup requirements
- Uptime

Figure 8.1 shows the typical DCP layout.

Machine 24 WORK CENTER DCP

Annual Work Days	247

TAKT Time		Per Shift Available Minutes =	480
Demand =	474110.4 pcs/day	Seconds per Shift =	28800
1 Shift TAKT	0.061 sec.	ADEQUATE TIME SCHEDULED	
		Hrs Scheduled per Day	16
2 Shift TAKT	0.121 sec.	Work Days For The Week	5
3 Shift TAKT	0.182 sec.	Uptime Factor	0.84

Part Number	Job I/D	Machine Number	Material Grade	Material Width	Material Thickness	ABC Code	Annual Usage	Cont. Quantity From Mch 24	Average Daily Usage	Average Daily Production Demand	Set-Up Time (minutes)	Die Number	Machine Cycle Time (minutes)	Pcs./hr (Machine Run Time Less Set-Up)	Strokes per Minute	Parts per Cycle	Cycle Time per Unit in Minutes	Run Time per Day Required in Minutes To Make ADU	Takt Image in Minutes	Available Time in Minutes per Work Day	EPEI in Days
										474110.4	80.25							307.66	0.0017	806.4	1
PW819		24	1008S	4.375	0.063	A	187899491	760708	76,070.8	76071	1.5	8.52	0.4594	92150	131	14	0.033	41.6			1
10FM62263BRZ		24	1008S	4.375	0.063	A	8628911	100000	34,934.9	34935	0.75	17.52	0.4417	68464	136	10	0.044	25.7			1
PWM83		24	1008S	4.125	0.066	A	40307691	100000	163,189.0	163189	1.5	9.52	0.4541	79913	132	12	0.038	102.9			1
PWM103		24	1008S	4.375	0.075	A	7634913	100000	30,910.6	30911	1.5	17.51	0.4507	33549	133	5	0.090	46.4			1
10FM62468		24	1010S	4.75	0.068	A	57399540	100000	23,237.0	23237	15	19.82	0.4520	66899	133	10	0.045	17.5			1
BM61363		24	1050S	4.08	0.063	A	12554085	100000	50,826.3	50826	15	5.2	0.4269	113351	141	16	0.027	22.6			1
BM61263		24	1050S	4.08	0.063	A	12533486	100000	50,742.9	50743	15	5.2	0.4385	110338	137	16	0.027	23.2			1
HTFM61763SF		24	1050S	4.08	0.063	A	5778808	100000	23,396.0	23396	15	9.52	0.4611	78701	130	12	0.038	15.0			1
HTBM61763		24	1050S	4.08	0.063	A	5138337	100000	20,803.0	20803	15	9.52	0.4399	82495	136	12	0.037	12.7			1

Figure 8.1 Dynamic capacity planner.

Role of Production Control

Planning from the PFEP will allow you to go beyond the boundaries of your MRP system, allowing finite scheduling and planning of your processes within the value stream. When you start looking at how to plan the value stream, the biggest gain you can make is to assign a value stream planner. This position will be the key to understanding all aspects of the processes within a value stream. Remember, the planner plans and the manufacturing processes execute that plan as laid out by the planner—to the letter. Any deviation from the plan has to be coordinated with the planner to ensure that the customer due date is met. When the plan is not met, that portion of the plan has to be moved to the following shift or day for completion before any new items can be added for the following shift or day. The planner understands the value stream, has access to forecasting tools, knows the precise committal date to the customer via the master scheduler, understands the lead time of internal and external suppliers, knows the capacity of the internal processes, follows the behind condition, and so on, so the manufacturing role is to execute the plan, not to control production.

This is a very important role in the Lean supply chain. It is not, however, recommended to have a value stream planner who assumes the role as the buyer for that value stream. Planners plan an area or group of machines. It is rare that they see the entire supply chain. That is why we still need material buyers, so that they can pull demand from all areas and make one consolidated buy. The buyer needs to be focused on the replenishment, forecasting requirements, supplier development, performance ratings, and the like. Buying and planning are both functions of a production control group, but because they see different levels of the value stream, they need to be separate functions.

Physically, the value stream production control people should be located in the same work area. For example, if you choose to have a buyer support two value stream planners, then the workspace layout would look like Figure 8.2.

Before you can just jump into planning the value stream from the PFEP, you are going to have to have the following elements of Lean in place:

■ Stability in the processes of the value stream.
■ Standard work at every process being properly followed.
■ Reliability in the processes through TPM activities, not predictability. (*Predictability* means that you can predict you'll get 100 pieces off the

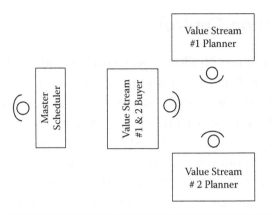

Figure 8.2 Planning layout.

screw machine, but you can't guarantee all 100 will be good. *Reliability* means you know you can get 100 pieces off the screw machine and can rely on all 100 to be good parts.)

■ Understanding that you need to plan to finite (true) capacity and not infinite capacity.

■ Overall Equipment Effectiveness (OEE) is measured and used to problem solve on all "Class A" machines.

■ Lean metrics are in place and used to gauge the state of the business or value stream.

■ Timed delivery routes are in place, utilizing kanban and supermarket locations.

■ Managing to the true customer due date, not the newly established customer promise date because you are behind.

■ Understanding the importance of the behind condition and having a plan to recover.

■ Manufacturing processes do not drive the business, but material and supply chain activities are the drivers (big change in thinking here).

Figure 8.3 is from an unknown author but is a very wise statement and viewpoint of the overall manufacturing process.

And the number one element, when you set a cell, line, or area up and you have determined that the area requires three people to run it based on the customer demand/takt time, then you set all the other aspects of Lean material movement into play: kanban, both internal and external; supermarkets; and time delivery routes will all work together.

> *Manufacturing:*
>
> *A series of material handling motions and information transactions,*
>
> *occasionally interrupted by a few seconds of value added work!*
>
> *Unknown but very wise!*

Figure 8.3 Manufacturing.

The biggest mistake you can ever make is to get the processes set up and working in a Lean manner and then something happens. Let's say you run out of a component for a cell; you can't build but now you have three people standing around. Because you are behind on work order #1873564, you decide to move these three people to Cell 20 to get another order out quicker. However, this won't work because you just added double the demand on the components for Cell 20 that was not planned for in the material flow. Cell 20 was set up for the rate and demand of requiring three people to meet a takt time of 60 units an hour, but they are now pulling at a rate of 120 units an hour. You had set your kanban at line 20 at 60 pieces an hour, and that is what the timed delivery route will bring you; the kanbans are not in the system to cover the newly created 120 pieces an hour.

By doing the above you have just covered the lowest portion of the cost in any unit you produce, the labor variance, and at the same time you sent the replenishment of the pull system into a tailspin. When you are using MRP to plan your shop floor and to forecast or order material through allocation, you can hide this behavior for a while; that is, until doing this catches you on the allocation portion of material flow.

There has to be a plan and it has to be adhered to. At one client site, they have completely shut off MRP except for the forecasting of raw materials to suppliers; we had to understand the capacity of the Pacemaker process in order for us to flow the components through the value stream in First-In First-Out (FIFO) order. This meant building a matrix of the machining process to the point that we understood:

- Actual overall equipment effectiveness percentage; understanding the importance of productivity, availability, and quality at the machine
- Measuring true first-time-through and what caused a rework, rerun, or return

Figure 8.4   Router example.

- Perimeter distance of every hole needing machining on every component from the design drawings
- Cutting inches per second off the machining tool
- Optimal speed per hole based on material
- Jog speed between holes
- Number of positions for each type of machining center; changeover order
- Assignment of primary and secondary machine designations by part number
- EPEI by primary machine
- Understanding Mean Time to Failure (MTTF) for cutting tools used at each process
- Using the PFEP to pivot table the level two components into the plan

All this just to machine 11 holes into a piece of PVC handrail (Figure 8.4) that's just one of three components, but we still have to get the bottom rail and spindles. The main thing this allowed us to do was to plan by machine; there had never been a correlation to the primary and secondary machining center and the demand versus capacity of each. The Bill of Materials (BOM) required far more capacity on certain machines and none on others; thus, the mix on the finished part number through the area was always causing overburden, people shuffling, and missed shipments.

Your first thought might be to place a supermarket between the machining centers and the assembly cells, but when you consider that there are 1,400+ component variations, we have to flow from the Pacemaker, not the machining center, to keep inventory down. Understanding the flow and process, the newly appointed value stream planner could evaluate the demand upon the capacity available, match the flow of each component across the correct primary or secondary machine, coordinate the pick/kitting of required purchased components, and issue a plan for each of the six areas, ensuring that the components all ended up in the particular cell's FIFO (next) lane on time and with the correct components needed to meet the customer's initial due date.

Removing the planning from the floor supervisor and placing it in the hands of a value stream planner who has all the data to make a good decision based on flow, the area went to a positive variance and remains there, something it had not been able to accomplish for many years.

Planning, discipline to maintain the plan, and understanding problem solving when the plan is not met are vital to the success of your Lean journey. Allowing production to change the plan to absorb labor will result in overproduction and building the wrong parts to meet customer demand. In addition, you will consume components that are required to build the products that your customer actually needs. Anyone can read a book, set up a cell, and balance the work content to takt time, but the real success is in how we plan, manage, and flow material.

Budgeting

From the above we have seen how to determine the method of delivery, containerization, and calculating personnel to support the delivery of parts. Manpower budgeting is the key here; a material handler (an associate, not an indirect person) should be the brightest and best of the associates you have in the facility. You have standard work at the process, good cell design, and a team leader at the cell for trouble-shooting, so let's work through the hiring process.

If you have a new hire for the production line (a direct person), you probably bring her into a job that has good work instruction, a defined workspace of a few feet, and a team leader there to jump in and support at a second's notice to keep the process flowing. The new hire doesn't have to be skilled, nor does she have to make decisions that affect the work area. But, when it comes to material handling (an indirect person), you bring in a temporary employee just to do "that material handling stuff" and throw them out on a timed delivery route that may or may not have work instruction, covers a large work area, and is supported by a team leader who is coordinating several people over the entire facility. We place her in an environment where she has to think for herself, ask her to make good rational decisions on a second's notice, place the fate of a dependent process on her shoulders, and provide very little close support. The material handler is the lifeblood of your organization, so she must have a deep understanding of the processes she supports.

To strengthen this premise of material importance, at Toyota, before you can become president, you have to spend time assigned to the production control department, meaning your desk is in the middle of the production control area, not in a plush front office.

Making a successful transition to Lean may require more than just focusing on the shop floor, it may require you to reconfigure your supporting processes. Understanding the needs and roles of production control, purchasing, and manufacturing is key to successfully implementing a Lean transformation.

Chapter 9

Supply Chain Complexity

In these days of global supply chains it is becoming increasingly common to outsource products with high labor content to so-called low-cost countries. Competition from low-wage countries across Asia and South America means that supply chains have become increasingly complex. Long lead times, scarcity of ocean containers, and logistics interruptions cause increasing inventory buffers to cover logistics performance variability. Global sourcing and low-cost competition put pressure on supply chains to reduce cost and improve response time to the changing needs of customers and suppliers. Traditionally, companies look to lower-cost sources of materials to reduce cost while missing opportunities within existing supply chains.

Supply Chain Integration

According to the *Manufacturing 2007 Executive Summary*, conducted by Advanced Manufacturing and the Manufacturing Performance Institute, almost one third of the manufacturing respondents have no supply chain integration with customers or suppliers.

> Where no integration exists, relationships are more likely based solely on transactions, which all but removes any ability to enhance value in the supply-chain for both suppliers and customers. As manufacturers optimize their internal operations, they increasingly look to their supply-chain relationships as the next target for cycle-time improvements and savings. Yet those manufacturers

without integration are missing their best chance for supply-chain improvements.

(Manufacturing 2007 Executive Summary, MPI, 2007 p. 1)

Supply Chain Integration (SCI) is about building relationships based on mutual success instead of reducing transactional cost. SCI means that you have to go beyond providing a product; you need to create a sustainable competitive advantage with your customers. Instead of providing a part, you need to provide a solution that may include inventory management, logistics services, a total component solution instead of a part, or operational expertise or a package of services that include some or all of the above. You want to make your relationship with your customer so deep that to change suppliers or "de-source" your products would require major modifications to your customer's processes or systems.

Successfully integrating your supply chain with customers and suppliers will make it more challenging for competitors to make inroads into your business. Your customer or supplier will no longer be looking at transactional costs; they will be looking at major disruptions to their business systems when evaluating a supplier change. Let's examine how we can use Plan for Every Part (PFEP) to evaluate supply chain relationships.

Inventory Impact

Low-cost sourcing is the buzzword of the millennium for North American companies. PFEP forces us to look at so-called low-cost sourcing beyond any piece price savings and look at the associated supply chain costs. Labor cost is normally 5–8% of a manufacturing process. Materials costs typically account for 40–70% of the cost of manufacturing. There is very little actual savings associated with low-cost sourcing where labor is a small portion of the cost of production. Any labor savings realized will be consumed by the cost of logistics in an extended supply chain. The answer is not low-cost countries but a low-cost supply chain or low-cost extended value stream.

In most horizontal organizations, purchasing professionals are rewarded by reducing material price. As we stated earlier, material costs make up a much higher percentage of unit costs than the cost of labor. If we look only at the unit costs, buying from low-cost countries makes economic sense. However, you must consider the supply chain costs. Let's examine the impact on our simple PFEP example when we change our supply chain for raw materials. Let's look at a hypothetical supply chain for a raw material (see Figure 9.1).

We have reduced the piece price by a few cents per piece, but our total raw materials inventory has gone from 10 days to 153 days. This does not count the finished goods in our warehouse or in our customer's warehouse. More important, now the pressure on manufacturing will go back to larger production batch sizes and reduced transportation frequency for finished goods, which will lead to larger safety stock levels for finished goods in the warehouse. Saving a few cents on a component leads us to make decisions that do nothing but grow larger inventories and increase carrying costs, not just for raw materials but across the entire supply chain.

It is not practical to purchase several weeks' worth of component parts from an overseas supplier in order to fill your ocean container, then bring those components inside your facility and put them in a storage location, just so you can run a Lean production system with frequent runs of small batches.

Component A Daily Usage 1000	Domestic (Daily delivery) (Inventory in Loop)	Domestic- Inventory in units	Asia Supplier (Weekly Delivery)	LCC - Inventory in units
Order lead time	2 days	2000	16 weeks	80000
Supplier Safety	1 day	1000	1 week	5000
Manufacturing Cycle	2 hours		10 hours	
Manufacturing Frequency	Daily	1000	Weekly	5000
Transportation Time	1 day	1000	8 Weeks	40000
Manufacturing Frequency (Finished Goods	Daily	1000	Daily	1000
Safety	(2 Days)	2000	4 weeks	20000
WIP	1 day	1000	1 day	1000
Finished Goods	1 day	1000	1 day	1000
Total Inventory	10 days	10000	153 days	153000

Figure 9.1 Supply chain for raw material.

Consider this example. We buy cast aluminum widgets from a supplier in India. We take the widgets and combine them with precision springs and gears to make wombats. We had a local supplier of widgets that delivered to us three times per week. We make 500 wombats per day for our customer. Our purchasing manager finds widgets in India that cost 8 cents less than our local supplier, a savings of 6%. But we have to buy a sea container of 10,000 widgets. Every time we buy a batch, we buy 20 days' worth of widgets. Now when we unload the container, we need more floor space to hold them. They don't fit into our normal storage location so we have to create an overflow or secondary location. We now have at least four moves for our material handlers instead of two, so we have doubled our material handling costs.

Now our warehouse supervisor is calling the planners asking how we can get rid of this excess inventory, so our planners and schedulers are looking at larger batch production runs. Because we don't have the space to store our large batches of finished wombats, we go to our customer to see if they will accept larger shipments at reduced frequency in exchange for a price concession. Now our planners and schedulers will revert to large batch production runs so they can get rid of the sea container of raw material that arrived last week from an overseas low-cost-country supplier. We then start giving price concessions to our customers to avoid holding larger amounts of finished goods. In the end, we may actually be making a lower gross margin than when we bought from the local supplier. This does not count the extra carrying costs or freight costs for the inventory and ocean freight!

The offshore suppliers have no understanding of your business or your customers. All through the summer of 2007 we heard one story after another of lead-tainted toys or other products coming out of Asia. The fact that these were discovered in North America in August and September indicates how long the supply chain is for holiday toys. It is very easy to see, by looking at the supply chain example above, the financial exposure a company has when you have a supply chain that is five or six months long. This also left the retailer with no option to replace the merchandise in time for the holiday shopping season, leading to lost sales. Any quality problems that occur with long-lead-time suppliers will only result in high-cost expedites or lost sales and customers. These additional costs rarely come back and get adjusted back into the piece price of the purchased part; instead they are accounted for as an additional cost variance for scrap or transportation expedite expense. Using this approach, your finance team never gets a true comparative piece price cost of the low-cost-country sourcing.

Toyota recognized long ago that a short, local supplier pipeline is superior to a long, centralized low-cost-country supplier pipeline. For that reason, Toyota encourages suppliers to build plants close to their manufacturing sites. This does not mean there is never a problem, but it does mean that the supply chain is infinitely easier to manage and recover if it is 200 miles long than if it is 20,000 miles long.

Yes, the purchasing professionals who negotiate contracts with offshore suppliers can negotiate quality clauses into a contract. With this, you have a good chance of recovering your component purchase price from the supplier (assuming they don't just close the door and go home). However, it is unlikely that you will recover the lost sales because your component is not available. And it is highly likely you will damage your reputation with your customers, by failing to provide them with a quality product, in the correct quantity, at the right time.

Logistics Cost

As anyone who has completed a transportation budget in 2010 knows, price volatility in transportation is a real problem. It is exaggerated when your supply chain is 6,000 miles long. When your supplier is that far away, you have no leverage or options when fuel prices go up 30%. In addition, you now become vulnerable to customs delays caused by tighter border security measures and the inelasticity of ocean and rail transportation capacity. And the one item that is taking a serious toll on supply chain logistics costing today, the devaluation of the U.S. dollar, is in many cases forcing companies to rethink low-cost countries and think more about onshoring, or bringing the work back.

Although it is impossible to buffer transportation volatility with any degree of reliability, it is possible to buffer transportation and delivery time volatility by adding inventory to your safety stock. In fact, this is the only way to protect service levels to your customers. If ocean freight times vary 10 days, you have no choice but to increase your safety stocks to cover this variability. If border crossing times can vary plus or minus one to three days, you have to add this buffer to your safety stocks.

With component supply chains 18 to 20 weeks long, you lose flexibility to react to changes in customer demand. Inasmuch as we rarely know where we want to have dinner 16 weeks from now, it is certainly unrealistic to assume that we know to a high degree of certainty what our customer wants to buy 16 weeks from now. This makes forecasting and ordering from long-lead-time suppliers an exercise in futility. The problem is that we try to design processes, create metrics, and then hold purchasing staff responsible for trying to manage with 98.5% accuracy in this environment while constantly changing the rules on inventory targets, turns, and levels.

We are not against responsible sourcing in today's global environment. In fact, I have seen many instances where it makes sense, especially if the item requires a high labor content and little or no technology. Fashion and textiles are high-labor-content items that are suitable for long supply chains. But where labor is a small percentage of the cost of the overall item and sufficient volume exists to justify building a local plant and investing in tools and equipment, you should try to build locally.

Other Supply Chain Considerations

We asked a manufacturing plant to put a map on the wall and put a pin on their suppliers' locations. The map revealed that the five largest suppliers were all located within five miles of the plant. Each supplier was delivering weekly using a different truck line. Our client went to the suppliers and made agreements for daily delivery using one carrier's milk run. The ability of the manufacturer to enter into long-term stable transportation contracts actually led to lower transportation costs and reduced inventories, from five days plus buffer to one day plus buffer. When completing your PFEP, it is important to identify your supply base and locations so you can engage in this kind of exercise. Your suppliers have the same problems you do: volatility of demand from your plant. Bringing the suppliers on board and making them part of your Lean transition will build loyalty within your supply base.

The supply chain complexity drawing in Figure 9.2 illustrates many of the factors that lead to volatility in our supply chain. It is critical to our success that we integrate all supply chain activities early on in our Lean transformation. As you involve purchasing, logistics, and packaging with manufacturing, you will find that these functions are usually willing to become engaged in your Lean transformation. Lacking a formal process of engagement, they build their silos to achieve their goals, lower piece price and lower transportation costs, without integrating their activities into the entire supply chain or Lean enterprise.

Supply chain complexity is more than distance; it includes quality considerations and lead time reduction as well as reduced variability. The

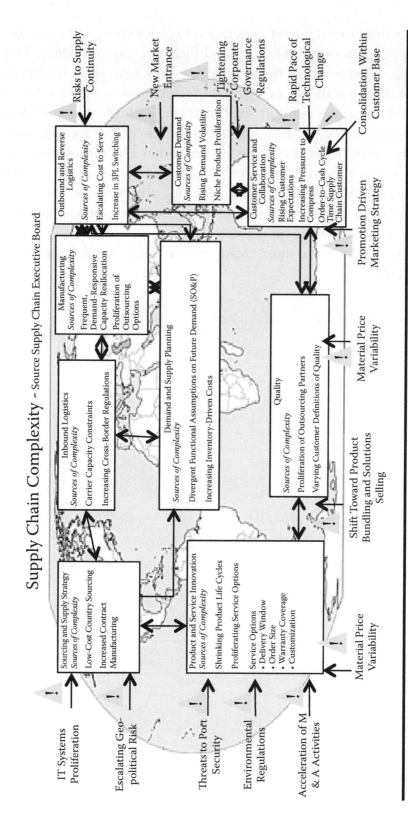

Figure 9.2 Supply chain complexity.

ugly reality is that over the next generation logistics costs will continue to increase as supplies of traditional fossil fuels decline and demand for the energy provided by these fuels increases. This is an issue of supply and demand. As the standard of living in Asia increases, demand for cars, electricity, and consumer comforts will increase, putting additional demand on resources that are being depleted. Transportation costs for long supply chains are the single highest risk we face as supply chain professionals. It is up to us to gather the data required to help our executives and supply chain professionals make the right decisions for our business. These decisions are not based on piece price; they are based on supply chain costs and creating a sustainable competitive advantage for us, our suppliers, and our customers. In short, we must create an integrated supply chain where we can all profit!

Chapter 10

Concluding Comments

Over the past 10 years, supply chains have become more complex. We are increasingly dependent on suppliers thousands of miles away and logistics systems operated by third parties. This does not mean that we should not work to remove waste from our supply chains. Plan for Every Part (PFEP) gives us the tools we need to understand the loops and cycle times that exist in our supply chains today. The same principles we use to establish inventory levels on our factory floor can be used to reduce supply chain cycle times in our global logistics networks.

Business today no longer has the luxury of adding inventory to cover long-distance supply chains. It is no longer possible or profitable to finance high levels of inventory even if it is carried on your balance sheet as an asset. The reality is that banks have reduced access to capital, making it difficult if not impossible to finance long supply chains. Now we must find ways to Lean out the supply chain by building only what the customer demands, buying and producing smaller lot sizes by utilizing more frequent deliveries. Turbo-charging your supply chain will not only reduce your dependence on borrowing but will actually return working capital to your business.

PFEP gives us the data we need to identify waste and reduce our supply chain cycle times. Our goal is to synchronize our supply chain so that our logistics moves at the same pace, the takt time, as our customer demand.

The authors have over 30 years' combined experience in supply chain, over 18 years of that with Toyota in North America. We have learned how to use PFEP to make operations efficient and effective. PFEP can help us understand the complexity of today's technology, engineering, and manufacturing demands that have been placed on the supply chain. The world was thought to be flat at one point in time and history has repeated itself, because with the speed at which we communicate and transact information, the world has once again become flat. And with that speed at which information flows and is transacted today, it is imperative that we change the way we flow, pull, move, and transact material to match the speed of information flow. PFEP will give you the vision to make your business competitive in the twenty-first century.

Changing to PFEP is not simple task, but it is a task any company can do, if you have the patience, dedication, and courage to take the first step.

Good luck on your Lean journey.

Robyn Rooks and Tim Conrad

References

Dennis, Pascal. 2007. *Getting the Right Things Done*. Lean Enterprise Institute

Krafcik, J. F. 1988. "Triumph of the Lean Production System," *Sloan Management Review, 30*(1), fall, pp. 41-52.

Rother, Mike and Shook, John. 1999. *Learning to See: Value Stream Mapping to Add Value and Eliminate MUDA*. Lean Enterprise Institute

Sorenson, Charles E. 1956. *My Forty Years with Ford*. New York: W. W. Norton

Index

Authors

Tim Conrad serves as Director of Operational Excellence for Gates Corporation, one of the world's leading manufacturers of industrial and automotive products, systems, and components, and a subsidiary of Tomkins PLC, a world-class global engineering and manufacturing group. Conrad oversees projects that link Gates Corporation's manufacturing plants and distribution centers with key customers.

Conrad served previously from September 2004 to August 2007 as Lean Implementation Manager of Gates World Wide Power Transmission operations. Prior to that Conrad spent nine years with Toyota Motor Manufacturing Kentucky, located in Georgetown, Kentucky. At Toyota, Conrad held positions in production planning, materials, and internal and external logistics.

Conrad holds a bachelor's degree from Northwood University in Midland, Michigan, and a master's in business administration with a specialty in international management from the University of Maryland.

Robyn Rooks is founder and president of MPnL Solutions, Inc., and has been helping companies in the United States, Canada, Mexico and Europe for the past 10 years to develop and implement Lean production systems with great success. He has created a culture change within nonunion and union organizations by working with management and the shop floor. He emphasizes building a "doing it with you" not a "doing it to you" culture.

Rooks started his Lean career in 1988 at Toyota Motor Manufacturing, Inc. (TMMK) in Georgetown,

Kentucky, where he was one of the original 1,700 employees hired to start the facility. Rooks spent two years on the production floor learning the material flow of TPS. He spent six years in the pilot "new model prototype" organization designing inter-departmental, external and internal assembly route delivery systems and balancing work content, writing standard work, designing supermarket layouts, setting kanban standards, ensuring flow and performing packaging approvals for new models.

Rooks spent four years in the production control and conveyance department, as a specialist, where he performed monthly planning for overseas suppliers, was North American engineering change implementation coordinator, was build-out and start-up coordinator, and as a North American parts ordering specialist was responsible for ordering components for the assembly lines. Rooks worked directly with suppliers to develop a Lean environment to support the needs of TMMK.